Stability and Decay of Heavy Nuclei

By

G. Naveya

Table of Content

CHAPTER 1

INTRODUCTION

Atomic nucleus, the building block of the universe is an enthralling many-body system held together by strong interactions. The evolution of our expanding cosmos, its structure, and the proton and neutron formation by weak interaction decoupling to subsequent alpha-cluster (^4He), exhibit the basic nuclear configuration and the beginning of matter formation. In fact, from the very moment of big-bang, fundamental forces have taken the major ground in governing the genesis and expansion of the universe (Aprahamian et al., 2005; Nazarewicz, 1998). Understanding a nucleus mainly lies in the detailed studies of its structure, which leads to a deeper understanding of the behaviour of strongly interacting matter ranging from ordinary nuclear matter to the dense cores of neutron stars, X-ray bursts, novae explosions, and thus to the Big bang - the theory of birth of the universe. In order to comprehend how nuclear processes shape the cosmos, and also to describe their astrophysical problems, knowledge of the nuclear properties of nuclei close to the stability region in the Segre chart is necessary for heavy-ion fusion, below the Fe-Ni region is crucial for stellar burning, the slow neutron capture process (s-process) and rapid neutron capture on the neutron-rich side of the chart (r-process), and the rapid proton capture along the proton drip line (rp-process) (Grawe et al., 2007). Also, nuclear temperature measurements, mass evaluation, binding energy, and half-life calculations of nuclei near the boundaries are an important characteristics for determining the stability of loosely bound nuclei and thereby aid in knowing about nuclear shape evolution near the boundary (Penionzhkevich, 2012). On this ground, nuclear physics helps to elucidate the typical processes in astrophysics. As a result, it assists in deciphering the features of nuclear processes that fuel stars and the production of new chemical elements. Almost all chemical elements heavier than lithium are generated in exploding and evolving stars, by stellar nucleosynthesis. In addition to sd shell and light nuclei formation in stellar events, the superheavy elements (SHE) are expected to be formed. Exploring the regions of superheavy may also lead to the expansion of the nuclear chart at the island of stability. In this context, the nuclei belonging to the superheavy region

(SHR) are being taken as the subject of research.

1.1 Heavy and superheavy nuclei

The quest to extend the periodic table of elements with nuclear charge and mass intrigued physicists' curiosity, and by 1935, elements upto Z = 92 were detected in nature. Two further heavy elements, Z = 93 & 94, were found in 1945. Beyond Z > 94 had been a missing piece of the puzzle for the formation of unique chemical elements, which do not occur naturally on Earth due to their short lifetime. However, their occurrence in space during specific celestial events such as supernova explosions, neutron star collisions, and other such happenings elucidate the element production and history of events in our solar system and the universe.

Superheavy nuclei (SHN) termed transactinide elements of Z $\geq$ 104, are formed in astronomical events that are not found on Earth, but they can be made in the laboratory by colliding stable light/heavy elements in hot or cold fusion reactions. SHN occupies the upper right corner of the nuclear landscape, whose extent is still unknown. Alternatively, SHE can also be identified by looking for their traces in extra-terrestrial sources, like the presence of Einsteinium and other actinide atoms in Przybylski star, etc., and their short life gives rise to a possibility that they were created in the decay of longer half-life closed-shell superheavy isotopes present in the atmosphere (Zhang and Wang, 2017).

Experimental studies reveal a breakthrough with the recent discoveries of SHN up to Z = 118. Discovery of new SHE with Z = 107 - 112 were synthesized at GSI (Gesellschaft für Schwerionenforschung) Germany with SHIP via cold fusion reaction evaporating 1 or 2 neutrons, and also a reaction with proper cross-bombardments of target ^{208}Pb and ^{209}Bi with projectiles $^{48}Ca,^{50}Ti,^{52,54}Cr,^{58}Fe,^{59}Co,^{64}Ni,^{65}Cu$ and ^{70}Zn (Khuyagbaatar et al., 2020; Hofmann and Münzenberg, 2000; Denisov and Sedykh, 2021). Element Z = 107, Bohorium was synthesized in 1981 from the reaction $^{54}Cr+^{209}Bi \rightarrow ^{263}Bh^* \rightarrow ^{262}Bh+1n$ giving rise to 5α-decay chains (Münzenberg et al., 1981). Later new isotopes ^{265}Bh, ^{266}Bh and ^{267}Bh were produced through hot fusion reaction in different labortories (Wilk et al., 2000; Eichler et al., 2000). Hassium, Z = 108 was synthesized in 1984 through $^{58}Fe+^{208}Pb$ reaction, successfully giving daughter nucleus $^{261}106$ which was obtained in the reaction $^{54}Cr+^{208}Pb$. Furthermore, three events of $^{265}108$ (Münzenberg et al., 1984) and one event of $^{264}108$ (Münzenberg et al., 1986) was observed and reconfirmed with target ^{107}Pb. Subsequent discovery on the isotopes of hassium were made through hot fusion reaction with $^{248}Cm+^{26}Mg$ (Dvorak et al., 2006) and $^{226}Ra+^{48}Ca$ (Oganessian et al., 2013) combinations. Also, the partial alpha decay half-life of ^{270}Hs with $T_\alpha = 7.6^{+4.9}_{-2.2}$ s was measured for the first time from experiments. Element Z= 109, Meitnerium was synthesized from $^{58}Fe+^{209}Bi$ reaction, in 1982. In this reaction, 1α-decay chain of ^{266}Mt was observed, decaying to previously synthesized ^{262}Bh element (Münzenberg et al., 1982). Z = 110 Darmstadtium, and Z = 111 Roentgenium, were discovered in 1994 individually from $^{62}Ni+^{208}Pb \rightarrow ^{269}Ds+1n$ (Hofmann et al., 1995b) and $^{64}Ni+^{209}Bi \rightarrow ^{273}Rg^*$ reaction respectively (Hofmann et al., 1995a). ^{281}Ds nuclei observed 13 α-decay chains with two half-lives of ≈ 1.8 and 65 ms, which was later confirmed by RIKEN observing 14 α-decay chains using the same target-projectile combination. Also, 3α-chains was observed for ^{272}Rg, which was later reconfirmed by RIKEN with 14 decay chains (Morita et al., 2004) and also by LBNL (Folden et al., 2004). Z = 112, Copernicium, was synthesized in 1996 at GSI with SHIP velocity filter through $^{70}Zn+^{208}Pb \rightarrow ^{278}112^*$ (Hofmann et al., 1996) reaction observing 1α-decay chain and a 2nd chain was detected in 2000 (Hofmann et al., 2002) in separate experiment conducted using same reaction set and thus reconfirmed later at RIKEN in 2004 (Morita et al., 2007).

Synthesis of elements with high number of neutrons was unsuccessful through cold fusion reaction. To extend the periodic elements further with higher mass nuclei, actinide targets are used in hot fusion reaction (evaporating 3 to 5 neutrons) leading to the discovery of elements Z = 113 - 118 at FLNR (Flerov Laboratory of Nuclear Reactions). Element Z = 113 developed at GSI in 1998 from $^{70}Zn+^{209}Bi \rightarrow ^{279}113^*$ reaction was reconfirmed at RIKEN in 2004 through hot fusion reaction. Elements Z = 114, 116 & 118 are synthesized at Flerov Laboratory of Nuclear Reactions (Dubna) and Lawrence Berkeley national laboratory (Berkeley) through hot fusion reactions. Z = 115 was synthesized at Joint Institute for Nuclear Research (JINR), Dubna in 2003 & Z = 117 was synthesized at JINR as a joint experiment with Oak Ridge National Laboratory (ORNL) in 2012. Synthesis of elements beyond Z > 118 ie., Z = 120 is presently under the process of synthesis at GSI Darmstadt.

Nuclei having closed proton and neutron shells are said to be doubly magic and such nuclei posses higher binding energy per nucleon (BE/A). In 1949, Nuclear shell model was introduced to discuss the stability of nuclei based on magic number of protons and neutrons that are equivalent to 2, 8, 20, 28, 50, 82 (Z & N) and 126 (for N). Around the doubly magic nucleus there exists a region of island of stability, since the neighbouring nuclides around this area are comparatively more stable than the farthest ones. The location of such magic islands in the superheavy region are an open problem to modern nuclear physics. To address this, several theoretical and experimental works on SHN have been performed since 1960. Among the advanced theoretical approaches using various phenomenological potential to study magic nuclei in the SHN region, the potential of modified oscillator type (Randrup et al., 1974) and Woods-Saxon model (Sobiczewski et al., 1966; Petrovich et al., 1976) has predicted Z=114 and N=184 as a spherical magic nucleus. The Skrymre- Hartree-Fock approach (SHF) (Ćwiok et al., 1996; Kruppa et al., 2000) and relativistic mean-field (RMF) models (Bender et al., 1999) predicted the possible shell closures beyond Z=114 at Z= 120, 126, 132, 138 and N= 184, 198, 228, 238, 258.

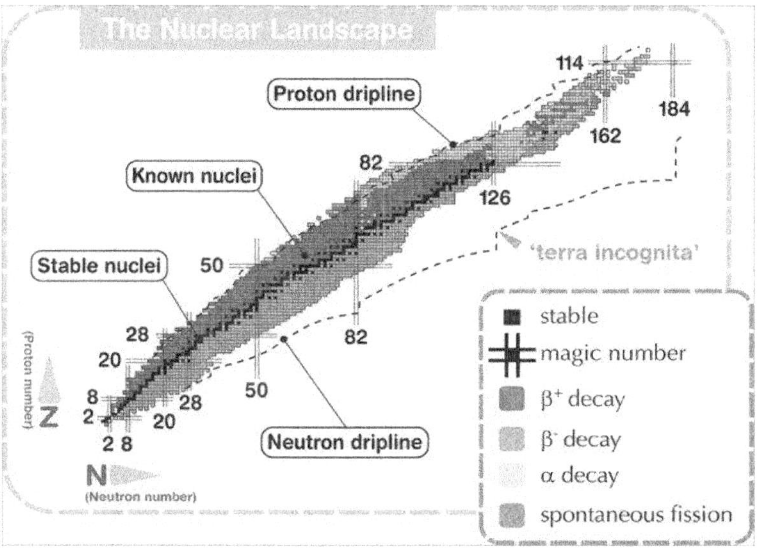

Figure 1.1: Nuclear Landscape. [Source : DOI:10.13140/RG.2.1.4573.3203, European ISOL Project (EURISOL)].

Similarly, the studies on shell corrections in the super and hyper-heavy regions by Denisov and Khudenko (Denisov and Khudenko, 2010) suggested the proton magic numbers, Z=114,

164, 210, 274, and 354; while the neutron magic numbers are N= 184, 228, 308, 406, 524, 644 and 772. The Nilsson-Strutinsky shell model (Nilsson et al., 1969; Mosel and Greiner, 1969) , Finite-Range Droplet model (FRDM) (W. D. Myers and Swiatecki, 1966; W. Myers and Swiatecki, 1967), and mean-field models have pointed out different possible magic numbers (Poenaru et al., 2002; Ćwiok et al., 1996; Ćwiok et al., 1999; Kruppa et al., 2000). Apart from this, there is evidence of subshell closures called the quasi-magic numbers; of which N=178 occurs near the most predicted N=184 closed-shell. Hence it is suggested that the proton shell closure and neutron shell closure beyond Z=82 and N=126 is at Z=114 and N=184 predicted in 1960, and it is expected that isotopes around this doubly magic nuclei will form an island of stability. All SHE synthesized in the laboratory are having few neutron numbers less to the predicted magic number N=184; for example, the heaviest isotope of Z=114 (^{292}Fl), is six neutrons less from N=184. But producing the isotopes approaching N=184 should have a considerable enhancement in its stability (Hamilton et al., 2013). A series of calculations based on the models developed were successful in reporting the well-known α-decay/cluster decay and SF half-lives.

1.2 Nuclear decay modes of heavy and superheavy mass regions

Radioactive decay is an important phenomenon associated with the nucleus, in which a nucleus undergoes transition via decay which can be alpha decay, beta decay, gamma decay, neutron emission, proton emission, SF, and cluster decay (fig. 1.1). Light and medium- heavy nuclei decay almost entirely via beta decay, electron capture, and proton emission. Heavy and SHN decay via beta decay, alpha decay, and SF, but beta decay for the SHN is slow as it proceeds via a weak interaction and is less favored compared to SF and alpha decay.

1.2.1 Alpha decay

A spontaneous process of transformation of the unstable radioactive nuclide into energetically stable one by emitting alpha particle, was first measured by Henry Becquerel in 1896 (Becquerel, 1896) and is one of the earliest radioactive decays to be found from the Uranium isotope. Its first experimental detection of deflection of radiations with magnetic field through artificial transmutation nuclear reaction using ^{14}N (α, p) in 1919, was performed by Rutherford (Rutherford,

1935). An alpha particle is structurally a Helium nucleus (^4_2He), consisting of 2 protons and 2 neutrons is favored over other such combinations primarily due to its high binding energy and small atomic mass. Alpha decay being one of the important decay modes in the chart of nuclides, its decay mechanism via quantum mechanical tunneling are crucial for nuclear physics research. Over more than 400 nuclei spanning across medium, heavy and super-heavy mass regions are prone to this decay, and is said to be dominant in heavy & superheavy region.

Interestingly in every radioactive decay, energy balance and probability of transition are its key attributes. A nucleus might undergo spontaneous decay when the alpha binding energy of such a nucleus becomes negative, ending in positive Q-value. The Q-value for alpha emission from nuclides across the nuclear landscape is shown in figure 1.2 and 1.3, here alpha emission are permitted in nuclides having positive Q-value and is forbidden in regions with Q-value being negative in a decay process. Systematic studies connecting α-particle decay energy and its mean lifetime was first deduced by Geiger and Nuttal in 1911 (Geiger and Nuttall, 1911). Following which Gamow (Gamow, 1928) & independently Condon and Gurney (Gurney and Condon, 1928) in 1928, discussed the theory of alpha decay process via tunneling which was the first theoretical dependence to Geiger and Nuttal and also as a first quantum mechanical application to tackle nuclear physics problems. The interplay of strong nuclear force and electromagnetic force governs the alpha decay process. The decay width (λ) primarily depends upon the penetration probability which is due to the alpha cluster penetration through the potential barrier. However, some models account for contributions from the preformation probability in addition to the penetration probability, and the preformation factor of alpha decay is found to not change significantly for most open-shell nuclei.

Experimental measurements of the alpha decay from medium and heavy isotopes have been primarily accomplished via induced spallation reactions or heavy-ion fusion evaporation reactions, followed by segregation using two approaches: (i) In-flight technique & (ii) Isotope separator On-Line (ISOL). In-flight separation is based on reaction kinematics and for this heavy ion beams are employed. The ISOL employs high energetic proton particles impinging on actinide. Studies employing these techniques from various In-flight separators such as SHIP at GSI (Germany), BGS (Berkeley)USA, GARIS/GARIS2 at RIKEN (Japan) (Düllmann, 2008) etc., and TRIUMF

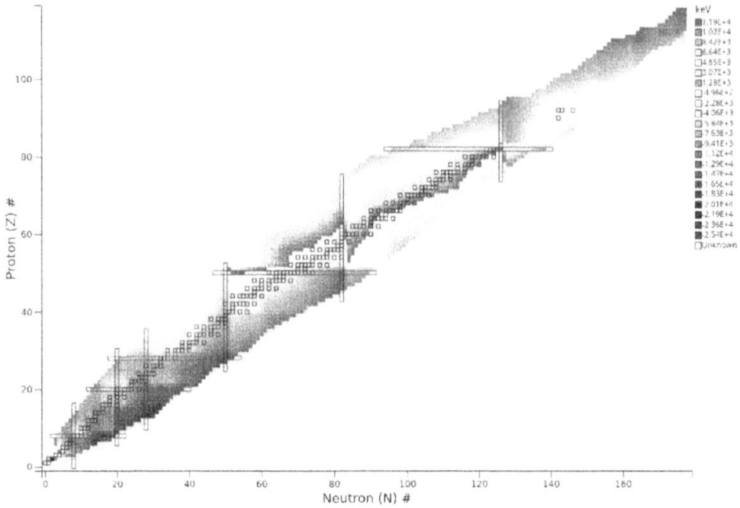

Figure 1.2: Q-value of alpha decay process in nuclear landscape. [Source: NuDat 3.0 decay data plotter of National Nuclear Data Center (NNDC), Brookhaven National Laboratory.]

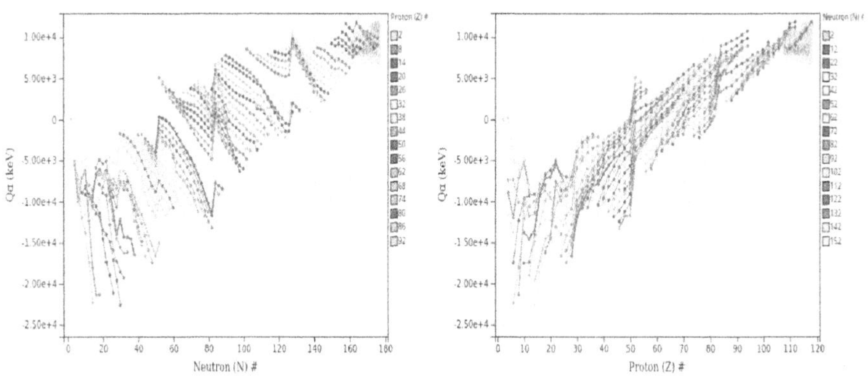

Figure 1.3: Q-value of alpha decay as a function of neutron number and proton number. [Source: NuDat 3.0 decay data plotter of National Nuclear Data Center (NNDC), Brookhaven National Laboratory.]

(Dilling et al., 2014) & ISOLDE (Borge and Blaum, 2018) techniques from ISOL based facilities substantially help to research studies on α decay and also to synthesize new elements and isotopes.

Nuclei with $A \geq 150$ and nuclei around doubly magic nucleus ^{100}Sn undergo alpha decay. So far ^{209}Bi (using scintillating bolometric techniques) is believed to be the longest stable alpha-emitting isotope with a half-life of 1.9×10^{19} years and decay energy of 3.137 MeV ever observed. Interestingly even the doubly magic nuclei ^{208}Pb with half-life $\approx 10^{123} years$ more than the age of our universe is deemed stable, inspite of it being unstable to alpha emission with alpha decay energy of 517.2 keV (Van Duppen and Andreyev, 2018).

Theoretical studies on alpha emitting heavy and superheavy regions, provide useful information about nuclear structure and stability. Several microscopic descriptions and phenomenological models play a forefront role in this regard. We list out some examples where alpha decay studies involving experimental observation and theoretical frameworks have been successful in revealing a wealth of information about the nuclear structure (Van Duppen and Andreyev, 2018) :

- **Super-allowed decay and clustering**

  The decay of ^{212}Pb points towards a configuration having two protons and two neutrons just outside the shell closures at Z=82 and N=126, the fastest decay, where the alpha particle clustering could be maximum and super allowed from this isotope. Since this decay involved proton and neutron from different shell model orbitals, it was suggested that decay of nuclei along N=Z line having same shell line to be the truly super-allowed. Thus decay property-related studies of nuclei close to ^{100}Sn and other nuclides can bring out information related to nuclear structure, clustering phenomena, and decay mechanism.

- **Shape Coexistence in the Neutron-Deficient Lead Region**

  Shape coexistence is a phenomenon exhibited by the atomic nucleus, wherein at very low excitation energies the atomic nucleus exhibits different nuclear shapes. Shape co-existence has been spotted in several regions in the nuclear chart. Alpha decay fine structure study of the neutron-deficient nuclei has helped in the identification and characterization of the low-lying 0^+ states. The alpha decay studies can uncover various phenomena and properties associated with the nuclear structure.

- **Properties of SHE and identification**

  Alpha decay energies can be measured accurately, and this supplement with the mass measurement of any one of the isotopes produced in α-decay chains originating from SHN can help in extracting accurate masses and binding energy of all the other isotopes involved in the chain. Also, the identification of the reaction products from the synthesis of SHN can be obtained with help of alpha particle-related measurements.

1.2.2 Cluster decay

The first observation of the emission of particles heavier than alpha particles was made by Rose and Jones (Rose and Jones, 1984) from the University of Oxford wherein ^{223}Ra isotope emitted a ^{14}C nucleus. In the year 1975–1976, de Carvalho et al.,(De Carvalho et al., 1974) reported such a rare radioactive decay process with calculations based on tunneling probability estimates from ^{238}U with clusters mass in the range $20 - 70$ and also indicated its shell effects were clearly associated with the decay rates, a process where either the emitted cluster or daughter has magic number nuclei. Cluster radioactivity is a rare process where the parent nucleus decays and emits a fragment which is heavier than the alpha particle but lighter than the fission fragment. The possible existence of such phenomena was established in the theoretical work of Sandulescu, Poenaru, and Greiner in 1980 (Sandulescu et al., 1980). Sandulescu et al., suggested the most favorable clusters such as, ^{48}Ca from Cf, Fm and No isotopes, ^{46}Ar from Pu and Cm isotopes, ^{32}Si & ^{34}Si from U isotope and ^{24}Ne & ^{28}Mg from Th isotopes. During the last three decades, many other clusters have been observed which include ^{20}O, ^{23}F, $^{24-26}Ne$, $^{28-30}Mg$, and $^{32,34}Si$, where the parent was either an actinide series nuclei or some heavier nuclei near to actinides (like Fr and Ra) (Bonnetti and Guglielmetti, 2007) (Ni and Ren, 2010).

Cluster radioactivity can either be thought of as strongly asymmetric fission or an exotic process of cluster formation inside the parent nucleus and then tunneling out crossing the barrier by assaults. The theoretical studies on exotic decays have inspired experimental groups to develop new methods for cluster identification and half-life measurements with a branching ratio of $10^{-16} - 10^{-9}$ with respect to the alpha decay. Fundamentally, the cluster decay is governed by shell closure effects of parent and daughter nuclei, The small half-life of cluster decay of a parent nucleus indicates the formation of a closed shell daughter and the long half-life indicates the stability of the parent itself.

But cluster decay from SHN is yet to be observed experimentally. In the recent years, the conception of cluster radioactivity has been redefined to accommodate the emission of particles with atomic number $Z_c > 28$ from the parent nuclei with $Z_p = Z_c + Z_d > 110$, where the daughter nuclei is either doubly magic ^{208}Pb or its close neighbours (Wei et al., 2017). Poenaru and Gherghescu theoretically investigated the $^{92,94}Sr$ cluster emissions from $^{300,302}120$ and have put forward the branching ratio values -0.10 and 0.49 relative to α-decay, which suggests that such cluster decay modes have good chances to be observed in competition with alpha decay (Poenaru and Gherghescu, 2018). Also a latest interesting experiment at the velocity filter SHIP (GSI Darmstadt) was carried out to produce the $^{299}120$ isotope in the fusion reaction $^{248}Cm(^{54}Cr, 3n)$ $^{299}120$. Hence in near future with the synthesis of $^{300,302}120$ isotopes, large clusters like $^{92,94}Sr$ can be expected to be observed in the decay.

1.2.3 Proton emission

A nucleus is said to be stable with the ratio between the number of neutrons and protons and any imbalance in this ratio will lead to radioactive decay via beta emission ie., transforming one neutron into a proton ($\beta-$ radioactivity) or a proton into a neutron($\beta+$ radioactivity). But, when the ratio exceeds the limit the nuclear forces can no longer play a role in binding the nucleons together and hence ejection of one or more nucleons takes place on the proton-rich side or neutron-rich side. Drip lines marked along the chart of nuclides (fig. 1.1) describe the limit of the number of protons and neutrons that can be bound within the nucleus and the rest leading to single or two proton emissions.

Analog to alpha decay, a new type of radioactive decay- 'Proton emission' is possible in the nuclei present beyond the proton drip line, and being unstable, they decay via proton emission. On the experimental front, Jackson et al., (Jackson et al., 1970) in 1970 made their first observation of single proton emission from isomeric state ^{53}Co to ground state ^{52}Fe. Since then several such observations were made from the ground and isomeric state for proton emission in the nuclei, $53 \leq Z \leq 83$. GSI Germany made its first observations in early 1980's. Though observation of one-proton emission was underway, studies on two-proton emission (2p) were trivial. 2p occurs when ^2He clusters are emitted simultaneously from the parent nucleus. Its first prediction was made in the year 1960 by Zel'dovich (Zel'dovich, 1960) and Goldansky (Goldansky, 1960, 1961). In 2002, results from experiments at GSI (Pfützner et al., 2002) and at GANIL (Giovinazzo et al., 2002) suggest that Fe-45 with excessive proton can simultaneously emit two protons at its ground state with a sufficiently longer half-life. Thereafter many such observation were made in detecting 2p emission from Zn-54 (Blank et al., 2000; Ascher et al., 2011), Mg-19 (Mukha et al., 2007) and other elements (Blank et al., 2005; Pomorski et al., 2011; Goigoux et al., 2016).

On the theoretical side, to study the proton radioactivity, a simple separation energy study reveals the possible 1p, and 2p emissions, eg., for negative separation energy below -1 MeV, 1p and 2p emission is possible with odd and even Z elements. Till date, 44 single proton decay events were detected, and much more is expected in near future to serve the understanding of nuclear deformation, masses, and structure of unstable nuclei. Also, with improvements in the theoretical and experimental front in the study of 2p emission, possible results from future observations may pave a way for potential studies relating to internal force governing the atomic nucleus and also beyond the proton drip line.

1.2.4 Spontaneous fission

Spontaneous fission (SF), another prominent decay mode apart from the α type is energetically feasible in the heavy and superheavy region, ie., $Z > 90$. It was first predicted by Bohr and Wheeler in 1939 (Bohr and Wheeler, 1939) and subsequently observed by Flerov and Petrzhak (Flerov and Petrzhak, 1940), in 1940. Since its discovery from ^{238}U, a number of transuranium nuclei with this type of radioactive decay have been reported in experiments. SF is seen from superheavy nuclei via α-decay chain, where the decay begins with alpha emission and terminates with SF, such is observed experimentally for Fl and Ts isotopes (Khuyagbaatar et al., 2014; Utyonkov et al., 2018). SF is also an important limiting factor that determines the stability of newly synthesized SHN and is the preferred decay mode in neutron-rich isotopes. SF of heavy elements in astrophysical studies of stars helps to determine the 'Fission recycling' in the r-process mechanism (Van Duppen and Andreyev, 2018). Theoretically, α-decay and SF share the same underlying mechanism in physics, i.e., the quantum tunneling effect. Till date among the 47 SHN synthesized via hot fusion reaction state 11 nuclei decay through SF (Hamilton et al., 2013).

The process of SF is more complicated to model as compared to other decay modes. The SF has been extensively studied using various theoretical approaches since its discovery (Poenaru and Gherghescu, 2014, 2016, 2017). The half-life calculation of SF involves complexity and in this regard, several empirical formulas were developed. Reliable empirical formulas for finding the half-life of SF using the parabolic potential approximation, taking nuclear structure effects into account have been proposed by Xu et al., (Xu et al., 2008; Xu and Ren, 2005). Also, empirical formulas developed by Ren and Xu (Ren and Xu, 2005) and Santhosh and Sahadevan (Santhosh and Sahadevan, 2010) have been widely used for finding the SF half-life.

Nuclei in the heavy and superheavy region ($84 \leq Z \leq 164$) are considered in this thesis for the study of structure, stability, and decay properties using cubic plus proximity potential, and the main attention is given to predicting proton and neutron magic numbers, which will be useful in locating the island of stability. Also, the different factors which impact the decay process are emphasized. This thesis is organized in a way that, **chapter 2** deals with the detailed description of various models which have been developed for decay half-life calculations, particularly the ones for proton emission, alpha emission, and heavy cluster emission. Various empirical formulas developed for proton emission, alpha decay, cluster emission, and SF are addressed. Also, a detailed description of the newly developed model, cubic plus proximity potential with improved transfer matrix method is being adopted in this thesis for studying the decay modes of heavy and superheavy nuclei, is presented. In **chapter3** the study on decay modes of even-even isotopes, α-decay chains for the yet-to-be-discovered SHE using the fission-like approach evaluated with the model being adopted is presented. **Chapter 4** deals with the influence of temperature on structural properties and decay behavior of nuclei. The effect of nuclear deformation and orientations on the decay properties is also addressed. The temperature and shape deformations are essential due to their key role in the synthesis and decay processes occurring in terrestrial and cosmic laboratories. In **chapter 5**, emphasis is given to the influence of screening effects on the half-life of alpha decay in the terrestrial and astrophysical environments that dictate the element formation and chemical abundance. This thesis work is been concluded in **chapter 6** with the scope for future studies.

References

Aprahamian, A., Langanke, K. and Wiescher, M. (2005). Nuclear structure aspects in nuclear astrophysics. *Progress in Particle and Nuclear Physics*, **54**, 535–613.

Ascher, P., Audirac, L., Adimi, N., Blank, B., Borcea, C., Brown, B., ... others (2011). Direct observation of two protons in the decay of ^{54}Zn. *Physical Review Letters*, **107**, 102502.

Becquerel, H. (1896). Sur les radiations émises par phosphorescence. *Comptes rendus*, **122**, 420–421.

Bender, M., Rutz, K., Reinhard, P.-G., Maruhn, J. A. and Greiner, W. (1999). Shell structure of superheavy nuclei in self-consistent mean-field models. *Physical Review C*, **60**, 034304.

Blank, B., Bey, A., Canchel, G., Dossat, C., Fleury, A., Giovinazzo, J., ... others (2005). First observation of ^{54}Zn and its decay by two-proton emission. *Physical review letters*, **94**, 232501.

Blank, B., Chartier, M., Czajkowski, S., Giovinazzo, J., Pravikoff, M., Thomas, J.-C., ... others (2000). Discovery of doubly magic ^{48}Ni. *Physical Review Letters*, **84**, 1116.

Bohr, N. and Wheeler, J. A. (1939). The mechanism of nuclear fission. *Physical Review*, **56**, 426.

Bonnetti, R. and Guglielmetti, A. (2007). Cluster radioactivity: An overview after twenty years. , **52**, 301-310.

Borge, M. J. and Blaum, K. (2018). Focus on exotic beams at isolde: A laboratory portrait. *Journal of Physics G: Nuclear and Particle Physics*, **45**, 010301.

De Carvalho, H., Martins, J., De Souza, I. and Tavares, O. (1974). *Spontaneous emission of heavy-ions from uranium* (Tech. Rep.). Centro Brasileiro de Pesquisas Fisicas.

Denisov, V. Y. and Khudenko, A. A. (2010). α decay of even-even superheavy elements. *Physical Review C*, **81**, 034613.

Denisov, V. Y. and Sedykh, I. Y. (2021). Production of super-heavy nuclei in cold fusion reactions. *Chinese Physics C*, **45**, 044106.

Dilling, J., Krücken, R. and Ball, G. (2014). Isac overview. *Hyperfine Interactions*, **225**, 1–8.

Düllmann, C. E. (2008). Physical separators for the heaviest elements. *Nuclear Instruments and Methods in Physics Research Section B: Beam Interactions with Materials and Atoms*, **266**, 4123–4130.

Dvorak, J., Brüchle, W., Chelnokov, M., Dressler, R., Düllmann, C. E., Eberhardt, K., ... Yeremin, A. (2006). Doubly magic nucleus $^{270}_{108}$Hs$_{162}$. *Physical Review Letters*, **97**, 242501.

Eichler, R., Bruchle, W., Dressler, R., Dullmann, C. E., Eichler, B., Gaggeler, H. W., ... Yakushev, A. B. (2000). Chemical characterization of bohrium (element 107). *Nature*, **407**, 63–65.

Flerov, G. N. and Petrzhak, K. (1940). Spontaneous fission of uranium. *Physical Review*, **58**, 275–279.

Folden, C. M., Gregorich, K. E., Düllmann, C. E., Mahmud, H., Pang, G. K., Schwantes, J. M., ... Hoffman, D. C. (2004). Development of an odd-Z-projectile reaction for heavy element synthesis: ^{208}Pb(^{64}Ni,n)^{271}Ds and ^{208}Pb(^{65}Cu,n)^{272}111. *Physical Review Letters*, **93**, 212702.

Gamow, G. A. (1928). Zur Quantentheorie des Atomkernes. *Zeitschrift für Physik*, **51**, 204-212.

Geiger, H. and Nuttall, J. (1911). Lvii. the ranges of the α-particles from various radioactive substances and a relation between range and period of transformation. *The London, Edinburgh, and Dublin Philosophical Magazine and Journal of Science*, **22**, 613-621.

Giovinazzo, J., Blank, B., Chartier, M., Czajkowski, S., Fleury, A., Jimenez, M. L., ... others (2002). Two-proton radioactivity of ^{45}Fe. *Physical review letters*, **89**, 102501.

Goigoux, T., Ascher, P., Blank, B., Gerbaux, M., Giovinazzo, J., Grévy, S., ... others (2016). Two-proton radioactivity of ^{67}Kr. *Physical review letters*, **117**, 162501.

Goldansky, V. (1960). On neutron-deficient isotopes of light nuclei and the phenomena of proton and two-proton radioactivity. *Nuclear Physics*, **19**, 482–495.

Goldansky, V. (1961). Two-proton radioactivity. *Nuclear Physics*, **27**, 648–664.

Grawe, H., Langanke, K. and Martínez-Pinedo, G. (2007). Nuclear structure and astrophysics. *Reports on Progress in Physics*, **70**, 1525.

Gurney, R. W. and Condon, E. U. (1928). Wave Mechanics and Radioactive Disintegration. *Nature*, **122**, 439.

Hamilton, J., Hofmann, S. and Oganessian, Y. T. (2013). Search for superheavy nuclei. *Annual Review of Nuclear and Particle Science*, **63**, 383-405.

Hofmann, S., Hessberger, F. P., Ackermann, D., Münzenberg, G., Antalic, S., Cagarda, P., ... Yeremin, A. V. (2002). New results on elements 111 and 112. *The European Physical Journal A - Hadrons and Nuclei*, **14**, 147-157.

Hofmann, S. and Münzenberg, G. (2000). The discovery of the heaviest elements. *Reviews of Modern Physics*, **72**, 733–767.

Hofmann, S., Ninov, V., Hessberger, F. P., Armbruster, P., Folger, H., Münzenberg, G., ... Leino, M. (1996). The new element 112. *Zeitschrift für Physik A Hadrons and Nuclei*, **354**, 229-230.

Hofmann, S., Ninov, V., Heßberger, F., Armbruster, P., Folger, H., Münzenberg, G., ... Leino, M. (1995a). The new element 111. *Zeitschrift für Physik A Hadrons and Nuclei*, **350**, 281-282.

Hofmann, S., Ninov, V., Heßberger, F., Armbruster, P., Folger, H., Münzenberg, G., ... Leino, M. (1995b). Production and decay of^{269}110. *Zeitschrift für Physik A Hadrons and Nuclei*, **350**, 277-280.

Ćwiok, S., Nazarewicz, W. and Heenen, P. H. (1999). Structure of odd-N superheavy elements. *Physical Review Letters*, **83**, 1108–1111.

Jackson, K., Cardinal, C., Evans, H., Jelley, N. and Cerny, J. (1970). 53com: A proton-unstable isomer. *Physics Letters B*, **33**, 281.

Khuyagbaatar, J., Mistry, A., Ackermann, D., Andersson, L.-L., Block, M., Brand, H., ... Yakusheva, V. (2020). The identification and confirmation of isomeric states in ^{254}Rf and ^{255}Rf through conversion electron detection. *Nuclear Physics A*, **994**, 121662.

Khuyagbaatar, J., Yakushev, A., Düllmann, C. E., Ackermann, D., Andersson, L.-L., Asai, M., ... others (2014). ^{48}Ca + ^{249}Bk fusion reaction leading to element Z= 117: Long-lived α-decaying ^{270}Db and discovery of 266Lr. *Physical review letters*, **112**, 172501.

Kruppa, A. T., Bender, M., Nazarewicz, W., Reinhard, P.-G., Vertse, T. and Ćwiok, S. (2000). Shell corrections of superheavy nuclei in self-consistent calculations. *Physical Review C*, **61**, 034313.

Morita, K., Morimoto, K., Kaji, D., Akiyama, T., Goto, S., Haba, H., ... Yoshida, A. (2007). Observation of second decay chain from ^{278}113. *Journal of the Physical Society of Japan*, **76**, 045001.

Morita, K., Morimoto, K., Kaji, D., Akiyama, T., Goto, S.-i., Haba, H., ... Zhao, Y. (2004). Experiment on the synthesis of element 113 in the reaction ^{209}Bi(^{70}Zn,n)^{278}113. *Journal of the Physical Society of Japan*, **73**, 2593-2596.

Mosel, U. and Greiner, W. (1969). On the stability of superheavy nuclei against fission. *Zeitschrift für Physik A Hadrons and nuclei*, **222**, 261–282.

Mukha, I., Sümmerer, K., Acosta, L., Alvarez, M., Casarejos, E., Chatillon, A., ... others (2007). Observation of two-proton radioactivity of ^{19}Mg by tracking the decay products. *Physical review letters*, **99**, 182501.

Myers, W. and Swiatecki, W. (1967). Anomalies in nuclear masses. *Arkiv för Fysik*, **36**, 343–352.

Myers, W. D. and Swiatecki, W. J. (1966). Nuclear masses and deformations. *Nuclear Physics A*, **81**, 1 - 60.

Münzenberg, G., Armbruster, P., Berthes, G., Folger, H., He, F., Hofmann, S., ... Hanelt, E. (1986). Evidence for ^{264}108, the heaviest known even-even isotope. *Zeitschrift für Physik A Atomic Nuclei*, **324**, 489-490.

Münzenberg, G., Armbruster, P., Folger, H., Heßberger, P., Hofmann, S., Keller, J., ... Hingmann, R. (1984). The identification of element 108. *Zeitschrift für Physik A Atoms and Nuclei*, **317**, 235-236.

Münzenberg, G., Armbruster, P., Heßberger, F., Hofmann, S., Poppensieker, K., Reisdorf, W., ... Vermeulen, D. (1982). Observation of one correlated α- decay in the reaction^{58}Fe on^{209}Bi $\rightarrow$ ^{267}109. *Zeitschrift für Physik A Hadrons and Nuclei*, **309**, 89-90.

Münzenberg, G., Hofmann, S., Heßberger, F., Reisdorf, W., Schmidt, K., Schneider, J., ... Thuma, B. (1981). Identification of element 107 by α correlation chains. *Zeitschrift für Physik A Atoms and nuclei*, **300**, 107-108.

Nazarewicz, W. (1998). Frontiers of nuclear structure. *Nuclear Physics A*, **630**, 239–256.

Ni, D. and Ren, Z. (2010). New approach for α-decay calculations of deformed nuclei. *Physical Review C*, **81**, 064318.

Nilsson, S. G., Thompson, S. and Tsang, C. (1969). Stability of superheavy nuclei and their possible occurrence in nature. *Physics Letters B*, **28**, 458–461.

Oganessian, Y. T., Utyonkov, V. K., Abdullin, F. S., Dmitriev, S. N., Graeger, R., Henderson, R. A., ... Yakushev, A. (2013). Synthesis and study of decay properties of the doubly magic nucleus ^{270}Hs in the ^{226}Ra + ^{48}Ca reaction. *Physical Review C*, **87**, 034605.

Penionzhkevich, Y. E. (2012). Exotic nuclei in astrophysics. *Physics of Particles and Nuclei*, **43**, 452–473.

Petrovich, F., Philpott, R., Robson, D., Bevelacqua, J., Golin, M. and Stanley, D. (1976). Comments on primordial superheavy elements. *Physical Review Letters*, **37**, 558.

Pfützner, M., Badura, E., Bingham, C., Blank, B., Chartier, M., Geissel, H., ... others (2002). First evidence for the two-proton decay of ^{45}Fe. *The European Physical Journal A-Hadrons and Nuclei*, **14**, 279–285.

Poenaru, D. N. and Gherghescu, R. A. (2014). Fission decay of ^{282}Cn studied using cranking inertia. *Journal of Physics G: Nuclear and Particle Physics*, **41**, 125104.

Poenaru, D. N. and Gherghescu, R. A. (2016). Spontaneous fission of the superheavy nucleus ^{286}Fl. *Physical Review C*, **94**, 014309.

Poenaru, D. N. and Gherghescu, R. A. (2017). Cluster preformation at the nuclear surface in cold fission. *Europhysics Letters*, **118**, 22001.

Poenaru, D. N. and Gherghescu, R. A. (2018). Alpha decay and cluster radioactivity of super heavy nuclei ^{303,304}120. *Europhysics Letters*, **124**, 52001.

Poenaru, D. N., Nagame, Y., Gherghescu, R. A. and Greiner, W. (2002). Systematics of cluster decay modes. *Physical Review C*, **65**, 054308.

Pomorski, M., Pfützner, M., Dominik, W., Grzywacz, R., Baumann, T., Berryman, J., ... others (2011). First observation of two-proton radioactivity in ^{48}Ni. *Physical Review C*, **83**, 061303.

Randrup, J., Larsson, S., Möller, P., Sobiczewski, A. and Lukasaik, A. (1974). Theoretical estimates of spontaneous-fission half-lives for superheavy elements based on the modified-oscillator model. *Physica Scripta*, **10**, 60.

Ren, Z. and Xu, C. (2005). Spontaneous fission half-lives of heavy nuclei in ground state and in isomeric state. *Nuclear Physics A*, **759**, 64–78.

Rose, H. J. and Jones, G. A. (1984). A new kind of natural radioactivity. *Nature*, **307**, 245-247.

Rutherford, L. (1935). Radioactivity: Old and new. *Nature*, **135**, 289–292.

Sandulescu, A., Poenaru, D. and Greiner, W. (1980). New type of decay of heavy nuclei intermediate between fission and. cap alpha. decay. *Soviet Journal of Nuclear Physics*, **6**.

Santhosh, K. P. and Sahadevan, S. (2010). Systematic decay studies of even–even $^{132-138}$Nd, $^{144-158}$Gd, $^{176-196}$Hg and $^{192-198}$Pb isotopes. *Nuclear Physics A*, **847**, 42 - 65.

Sobiczewski, A., Gareev, F. and Kalinkin, B. (1966). Closed shells for Z > 82 and N > 126 in a diffuse potential well. *Physics Letters*, **22**, 500–502.

Utyonkov, V., Brewer, N. T., Oganessian, Y. T., Rykaczewski, K. P., Abdullin, F. S., Dmitriev, S., ... others (2018). Neutron-deficient superheavy nuclei obtained in the $^{240}Pu + ^{48}Ca$ reaction. *Physical Review C*, **97**, 014320.

Van Duppen, P. and Andreyev, A. N. (2018). Alpha decay and beta-delayed fission: Tools for nuclear physics studies. In C. Scheidenberger and M. Pfützner (Eds.), *The euroschool on exotic beams - vol. 5* (pp. 65–116). Cham: Springer International Publishing.

Wei, K., Zhang, H. et al. (2017). Cluster preformation law for heavy and superheavy nuclei. *Physical Review C*, **96**, 021601.

Wilk, P. A., Gregorich, K. E., Türler, A., Laue, C. A., Eichler, R., Ninov, V., ... Hoffman, D. C. (2000). Evidence for new isotopes of element 107: ^{266}Bh and ^{267}Bh. *Physical Review Letters*, **85**, 2697–2700.

Xu, C. and Ren, Z. (2005). Systematical law of spontaneous fission half-lives of heavy nuclei. *Physical Review C*, **71**, 014309.

Xu, C., Ren, Z. and Guo, Y. (2008). Competition between α decay and spontaneous fission for heavy and superheavy nuclei. *Physical Review C*, **78**, 044329.

Zel'dovich, Y. B. (1960). JET 38 (1960) 1123. *Soviet Physics*, **11**, 812.

Zhang, Y. and Wang, Y. (2017). Systematic study on the competition between α-decay and spontaneous fission of superheavy nuclei. *Nuclear Physics A*, **966**, 102-112.

Ćwiok, S., Dobaczewski, J., Heenen, P.-H., Magierski, P. and Nazarewicz, W. (1996). Shell structure of the superheavy elements. *Nuclear Physics A*, **611**, 211-246.

CHAPTER 2

METHODOLOGY

2.1 Introduction

Radioactive decay in nuclear physics has grasped a great sign of interest since its inception in 1896 by Antonio Henri Becquerel, which played a significant role in solving the major puzzle to the birth of the universe. In general, an unstable nucleus becomes stable one through decay ie., the instability caused by mass excess leads to α, β or γ decay depending upon the number of nucleons, and the neutron to proton ratio. If the number of neutrons is very high the feasibility of α-emission is high ie., when the neutron to proton ratio exceeds 1.4:1. Releasing the excess energy in the form of γ-quanta is associated with the γ decay where the state of the energy level is altered, while the charge number of nucleus is altered in the β decay. Interestingly in the year 1984 Rose and Jones (Rose and Jones, 1984) detected the light ion (^{14}C) emission from ^{223}Ra isotope which was confirmed by Aleksandrov et al., (Aleksandrov et al., 1984), Gales et al., (Gales et al., 1984) and Price et al., (Price et al., 1985) independently, called cluster radioactivity. Elucidating the phenomenon of such a physical system and to describe its decay processes, a well established theoretical formulation is vital. Ever since, different theoretical models and experiments by various groups of theoreticians and experimenters have been developed and calculations have been performed from macroscopic to microscopic range. Theoretical models which correlate the experimental results to satisfy the phenomenological perseverance is highly essential in nuclear studies. Fission being the complex process in nuclear decay it may be well understood by considering the structure, stability and forces of strongly interacting nuclear matter. Inspite of the vast knowledge in such a process, it is still unclear to define a common suitable model and hence constructing simplified unified models and making nuances based on observations will illuminate new sights into the real behaviour of nuclei and also provides a foundation for further development towards making a complete theory. Keeping in track with the available literature, the methodology followed in this thesis is enumerated in the following sections: section 2.2 reviews different theoretical models developed over years, and the tunneling probability method such as WKB and Transfer matrix method are reviewed in section 2.3. The improved version of the transfer matrix method with cubic plus proximity potential used in this thesis work is presented in section 2.4. The inclusion of temperature and deformation in the potential is addressed in the respective chapters.

2.2 Review of theoretical models

Theoretical models developed based on the suitable mathematical formulation play a key role in developing technology dealing about a nucleus, especially the radioactive decay or emission in the process of reaching nuclear stability. As such the proton emission, alpha decay, and cluster decay were to be treated as a fission-like process since a parent nucleus gets split into a daughter nucleus and a fragment. Physics behind the fission model can be delineated using two theoretical approaches such as Unified Fission Model (UFM) and Preformed Cluster Model (PCM). In UFM, the fragment gets formed while it is detaching away from the parent nucleus and tunnels through the barrier before it reaches the saddle point. In PCM, the cluster is said to be preformed within the parent nucleus before it penetrates through the interaction barrier. A detailed description of these two models are given below.

2.2.1 Unified Fission Model

In UFM, α-particle as well as the heavier clusters have equal probability/ fair possibility of being preformed. The phenomenon of alpha decay, heavy cluster decay and spontaneous fission in UFM is treated as an equivalent process, with dissimilarity in degree of symmetry namely Super-asymmetric, asymmetric and symmetric fragments. Super-asymmetry and asymmetric fragments are formed in an alpha and cluster decay process, where the difference in size of nuclei is large, if Gamow theory of α-decay is followed and symmetric fragments are generally formed via Spontaneous fission (SF). These three possibilities of fragment formations are based on two approaches- (i) whether the fragment is getting formed within the parent nucleus, and (ii) alternatively the parent nuclei is assumed to keep deforming continuously to form fragments until the saddle point configuration is reached, which is called as the Unified Fission Model. In this fission model there are two important regions (i) The Pre-scission or the overlapping region where no preformation factor is taken into consideration. The Analytical super-assymmetric fission model (ASAFM) by Poenaru et al., (D. Poenaru et al., 1979), approximated the overlapping region to second order polynomial, Shanmugam and Kamalakaran (Shanmugam and Kamalaharan, 1988) used third order polynomial in finite range Yukawa plus exponential potential to fix the top of the barrier and shape of the potential as suggested by Nix (Nix, 1969). Shi and Swiatecki (Shi and Swiatecki, 1985a,b) used simple power law interpolation for their calculation. (ii) In the post scission region which is a sum of Coulomb and centrifugal potential, the half life ($T_{1/2}$) for the emitted fragment can be found using different interaction potentials for calculating the barrier penetration probability P (Gupta and Greiner, 1994). Hence,

$$P = \exp[-2\int_{R_1}^{R_2}\sqrt{(2\mu\hbar^2)(V-Q)dR}], \tag{2.1}$$

where R_1 and R_2 are the turning points at which the sum of nuclear and Coulomb potential (V) is equal to the energy released during the decay process (Q) ie., $V = Q$, and μ is the reduced mass. The assault frequency ν in terms of Zero-point vibration energy E_v, is given by

$$\nu = \frac{2E_v}{h}, \tag{2.2}$$

where E_v is an empirical formula framed by fitting experimental data of several alpha and cluster emitters

$$E_v = Q[0.056 + 0.039\exp\frac{(4-A_2)}{2.5}] \quad Q > 0;\ A_2 \geq 4, \tag{2.3}$$

where A_2 is the daughter nucleus. The α-decay half life using the traditional WKB approximation method is given by

$$T_{1/2} = \frac{\ln 2}{\nu P}. \tag{2.4}$$

This model is being used in this thesis with substitution of Cubic plus proximity potential.

2.2.2 Preformed Cluster Model

The clusters are said to be performed within the parent nucleus well before it penetrates through the barrier in the preformed cluster model. The preformed cluster model is of two stage processes ie., (i) formation of clusters well inside the parent nuclei described through clustering theory, or fragmentation theory or empirical estimates, and (ii) the penetration process. Hence, the decay constant in PCM with the preformation factor P_0 associated with the contribution of different emitted clusters is,

$$\lambda = \nu P P_0. \tag{2.5}$$

Here P is the tunneling probability of the outgoing fragments, P_0 is the probability at which the cluster gets preformed within the parent nucleus. Clusters that are within the parent nucleus of different sizes have different preformation probabilities. The preformation probability can be obtained via different approaches, such as, using the shell model and the other collective model descriptions of nuclei. Also, preformation probability can be extracted from the experiments. This could also be referred to as a spectroscopic factor. We briefly describe some preformation probabilities obtained by different formalisms. The preformation factor for the proton can be taken as, $P_0^{proton} = 1$ since the proton is already present in its form within the nucleus. The preformation factor used for alpha particle emission in different works is different. Some formalisms predict P_0^{α} to be about 10^{-2}, and the cluster preformation taken to scale as

$$P_0^{cluster} = (P_0^{\alpha})^{(A_2-1)/3}, \tag{2.6}$$

where A_2 is the emitted cluster's mass number (Zhang et al., 2009). Also, the tunneling probability through the interior barrier can be interpreted as preformation factor (D. N. Poenaru and Greiner, 1991). Some models obtained the preformation factor by analyzing the expression of the following form (Deng and Zhang, 2021),

$$P_0^{\alpha} = \frac{T_{1/2}(\text{theory})}{T_{1/2}(\text{experimental})}. \tag{2.7}$$

The α preformation factor using the Cluster Formation Model (Alsaif et al., 2020) is based on an interesting picture. It has recently been found that the daughter cluster can be considered inert and the clusterization state depends on the nucleons on the surface of the nucleus and so the alpha cluster formation was only included.

$$P_0^\alpha = \frac{\langle \Psi_F | H_F | \Psi_F \rangle}{\langle \Psi_F | H_F | \Psi_F \rangle + \langle \Psi_r | H_r | \Psi_r \rangle}. \tag{2.8}$$

where Ψ_F describes the alpha-cluster formed in the clusterization state, H_F is the Hamiltonian which represents internal interactions among the four nucleons, and Ψ_r describes alpha-cluster relative motion with respect to the core or daughter inside the parent nuclei. These two wavefunctions that describe the formation and the separation mechanisms can be obtained from the solution of the time-independent Schrodinger equations, $H_F\Psi_F = E_F\Psi_F$ and $H_r\Psi_r = E_r\Psi_r$. The total energy $E = E_F + E_r$. Thus in terms of energy it is shown that, $P_0^\alpha = \frac{E_F}{E}$, where

$$E_F = \begin{cases} 2S_p + 2S_n - S_\alpha & \text{(even-even)} \\ 2S_p + S_{2n} - S_\alpha & \text{(even-odd)} \\ S_{2p} + 2S_n - S_\alpha & \text{(odd-even)} \\ S_{2p} + S_{2n} - S_\alpha & \text{(odd-odd)} \end{cases}$$

where S_p is one-proton separation energy, S_n is one-neutron separation energy, S_{2p} is two-proton separation energy , S_{2n} is two-neutron separation energy, and S_α is alpha particle separation energy. The separation energies can be obtained from,

$$\begin{aligned} S_\alpha(A,Z) &= B(A,Z) - B(A-4,Z-2) \\ S_p(A,Z) &= B(A,Z) - B(A-1,Z-1) \\ S_n(A,Z) &= B(A,Z) - B(A-1,Z) \\ S_{2p}(A,Z) &= B(A,Z) - B(A-2,Z-2) \\ S_{2n}(A,Z) &= B(A,Z) - B(A-2,Z). \end{aligned} \tag{2.9}$$

Santhosh and Jose (K. P. Santhosh and Jose, 2021) proposed few formulas to calculate the preformation probability of cluster decay which depends on various quantities such as the size of cluster, product of cluster proton number and daughter nuclei etc. The Q value dependent preformation is

$$P_0^{cluster} = 10^{aQ+bQ^2+c}, \tag{2.10}$$

with the parameters $a = -0.25736$, $b = 6.37291 \times 10^{-4}$ and $c = 3.35106$. This provides preformation probability for light and heavy clusters in the heavy and superheavy parent nuclides.

In lieu of the above discussion, various other theoretical models based on these approaches play a vital role in nuclear structure studies, and few of them are presented in table 2.1.

Table 2.1: Some theoretical models employed in previous studies related to decay processes.

Model	UFM/PCM	Model detail
The Analytical Super Asymmetric Fission Model (ASAFM) (D. Poenaru et al., 1979)	UFM	An analytical relationship developed to evaluate half life for a system with charge asymmetry, shell corrections, angular momentum and excitation effects from the usual numerical super asymmetric method developed by Poenaru et al., in 1979. For the overlapping region a second order polynomial was used.
The Proximity Potential Model (Błocki et al., 1977)	UFM	Simple interpolation formula was used.
Cubic plus Yukawa-plus Exponential Potential Model (CYEM) (Shanmugam and Kamalaharan, 1988)	UFM	For the post scission region the potential is considered as a function of distance between centre of mass of fragments and for the overlapping region it is approximated to third order polynomial.
Coulomb and Proximity Potential Model (CPPM) (K. P. Santhosh et al., 2015)	UFM	The potential is sum of coulomb plus the proximity and the centrifugal term, for the post scission region. For the pre-scission region a general n-power law formula is used.
The Microscopic Model (Blendowske et al., 1987)	PCM	The potential used is sum of semi empirical formula for heavy ion, coulomb and centrifugal term.
Cosh potential model (Buck et al., 1992)	PCM	The square well potential is replaced by cosh form of geometry with nonzero diffuseness, and the surface charge coulomb potential is replaced by a form appropriate to a point charge α particle interacting with a uniformly charged spherical core.
The Double folded Michigan-3 Yukawa (M3Y) Potential Model. (Săndulescu et al., 1992)	PCM	Heavy ion emission including deformation effects was studied using M3Y potential.

2.3 Tunneling probability methods

Tunneling probability is one of the essential problems in quantum mechanics. To understand the role of tunneling for specific nuclear systems, the construction of simplified potential energy profiles are important and to investigate one must extract the transmission coefficient by solving the one-dimensional time-independent Schrodinger equation. Approximation methods with slowly varying potentials are well assigned to tackle such problems. Following subsections brief the traditional WKB method and Transfer matrix method.

2.3.1 WKB method

This method was developed in the year 1926, by Wentzel, Kramers and Brillouin and is named after them as WKB (Kramers, 1926). In quantum mechanics, WKB calculations are done in semi-classical manner in which the presumed Schrodinger wavefunction is an exponential function that is slowly varing with phase change and amplitude comparative to λ, the de-Broglie wavelength. To intuit the concept of approximation method to 1D problems, one-dimensional Schrodinger equation is considered. Then,

$$\frac{d^2\Psi}{dx^2}+k(x^2)\psi=0 \tag{2.11}$$

where

$$k(x)=\frac{2m}{\hbar^2}(E-V)^{1/2} \quad E>V(x); \quad k(x)=i\frac{2m}{\hbar^2}(V-E)^{1/2} \quad E<V(x). \tag{2.12}$$

If the value of k(x) is a constant then the solution to the function is

$$\psi=e^{\pm ikx}. \tag{2.13}$$

When k is no more a constant, the possible solution varies slowly with x depends on k is

$$\psi=e^{\pm i\int k(t)dt}. \tag{2.14}$$

Substituting equation 2.14 in 2.11, the Schrodinger equation takes the form,

$$\frac{d^2\Psi}{dx^2}+k(x^2)\psi=\pm ik^{'}(x)e^{\pm i\int k(t)dt}. \tag{2.15}$$

The solution 2.14 is arrived when k' is assumed to be zero and eqn. 2.15 remains as a good approximation when k' is considered to be negligible enough for $\mid k' \mid \ll k^2$. Assuming $u(x)$ to be a complex function, the solution to Schrodinger equation from equation 2.11 is $\psi(x)=e^{iu(x)}$ and substituting it in 1D equation, the normalised version of equation 2.11 will become

$$iu''(x)-u'(x)^2+k(x)^2=0. \tag{2.16}$$

since the approximation assumes the potential considered is slowly varying. Under this condition, a zeroth order approximation will be

$$u_0=\int_{x_0}^{x}k(t)dt, \tag{2.17}$$

and thus for an nth successive process the normalised version of Schrodinger equation can be

$$u_n=\pm\int_{x_0}^{x}\sqrt{k^2(t)+iu''_{n-1}(t)dt}. \tag{2.18}$$

Therefore one can write the 1st order approximation as

$$u_1=\pm\int_{x_0}^{x}\sqrt{k^2(t)+iu''_0(t)dt} \quad \approx\pm\int_{x_0}^{x}k(t)+\frac{i}{2}\frac{k'(t)}{k(t)}dt \tag{2.19}$$

$$u_1(x)=\int_{x_0}^{x}\pm k(t)dt+\frac{i}{2}lnk(x)+C.$$

Here $k'(t)<<k^2$ as defined earlier. The first order term will be quite close with zeroth order term and the wavefunction defined is called the WKB approximation with C being neglected under normalization condition, and so,

$$\psi_1(x)=\frac{1}{\sqrt{k(x)}}exp[\pm i\int_{x_0}^{x}k(t)]. \tag{2.20}$$

This when applied to the barrier potential problem, the probability of transmission can be evaluated using,

$$T \approx \frac{|F^2|}{|A^2|} = exp[-2\int_0^x k(t)dt]. \tag{2.21}$$

2.3.2 Transfer matrix method

In quantum physics, the probability flux carried by the transmitted wave in comparison to the incident wave is rendered well by the transmission coefficient. For calculating the tunneling probability, WKB approximation provides a simple formula that is easy to execute numerically and is reasonably accurate but few shortcomings seen in the regions where the potential changes abruptly and encounters difficulty when the incident energies are close to the barrier height, makes it hard to implement. To circumvent these constraints, numerical approaches that solve the transmission coefficient must be used and for this a step wise approximation method called the transfer matrix is administered. Numerical methods need a finite domain and boundary conditions for their study. Adhering this method solves one dimensional Schrodinger equation at every point forming two complex wave components and its derivatives are equated at the boundaries and hence the matrices at the boundary are determined. The results obtained through this method are satisfactory than WKB approximation (Biswas and Kumar, 2014).

The Schrodinger's wave equation with spatial coordinates x in 1D space is,

$$\frac{-\hbar^2}{2m}\frac{d^2\Psi}{dx^2} + V(x)\psi(x) = E\psi(x). \tag{2.22}$$

Schrodinger equation reduces to 1D Helmholtz equation within any region of space where V is constant, then

$$\psi = Ae^{ikx} + Be^{-ikx}; \quad k = \sqrt{\frac{2m[E - V(x)]}{\hbar^2}}. \tag{2.23}$$

The wave number k is real and positive when E > V and $e^{\pm ikx}$ represents either the right or left moving wave and k is purely imaginary when E < V, $e^{\pm ikx}$ represents exponential decreases towards right or left. Thus the above equation 2.23 can be rewritten as

$$\psi = \psi_+(x) + \psi_-(x). \tag{2.24}$$

Considering the constant potential function $V(x)$, as small regions of space having the values V_1, V_2, V_3..... and is applied to transfer matrix method. The wave components at position x_b can be found, when the wave components at point x_a are specified. These, in common, are connected via a linear relation as,

$$\psi_b = M(x_a, x_b)\psi_a. \tag{2.25}$$

where

$$\psi_b = \begin{bmatrix} \psi_+(x_b) \\ \psi_-(x_b) \end{bmatrix}, \quad \psi_a = \begin{bmatrix} \psi_+(x_a) \\ \psi_-(x_a) \end{bmatrix}, \tag{2.26}$$

and $M(x_a, x_b)$ is the 2×2 transfer matrix. The solution to the transfer matrix can be evaluated from the chosen potential and energy.

Transfer matrix for simple case

To Construct a transfer matrix for a simple potential case, the eqn. 2.23 may be considered in the region between x_a and x_b. Then at position x_a and x_b the wave components will be,

$$\psi_a = \begin{bmatrix} Ae^{ikx_a} \\ Be^{ikx_a} \end{bmatrix}, \quad \psi_b = \begin{bmatrix} Ae^{ikx_b} \\ Be^{ikx_b} \end{bmatrix}, \tag{2.27}$$

here each wave component ψ_b= $e^{ik(x_b - x_a)}$ ψ_a and equation 2.23 can be written in a linear relation by eliminating A and B as

$$\psi_b = M_0(k, x_b - x_a)\psi_a. \tag{2.28}$$

Here $L = x_b - x_a$ is the segment length and k is the wave number, the matrix $M_0(k, L)$ between regions x_a and x_b will be,

$$M_0(k, L) = \begin{bmatrix} e^{ikL} & 0 \\ 0 & e^{-ikL} \end{bmatrix}. \tag{2.29}$$

Similarly, for the next region, x_o lies in between x_a and x_b and assuming the potential step to be infinitesimally close to each other, then the wave-numbers for such a potential will be

$$k_\pm = \sqrt{\frac{2m[E - V_\pm]}{\hbar^2}}. \tag{2.30}$$

The quantum mechanical wave function must be continuous everywhere including at the point x_o. The derivative of the wavefunction should also be continuous at this point, and combining the schrodinger equation at these two conditions yield a single matrix equation. The inverse of the obtained single matrix will be,

$$\psi_b = M_s(k_+, k_-)\psi_a \tag{2.31}$$

where

$$M_s(k_+, k_-) = \frac{1}{2}\begin{bmatrix} 1+\frac{k_-}{k_+} & 1-\frac{k_-}{k_+} \\ 1-\frac{k_-}{k_+} & 1+\frac{k_-}{k_+} \end{bmatrix}. \tag{2.32}$$

Hence it is possible to derive the transfer matrix for any region of space in the potential region x_a and x_b. By left multiplying a sequence of transfer matrices one after the other yield two types of transfer matrices, such as M_0 and M_s representing matrix to cross a uniform segment and to cross potential step respectively as given in eqn. 2.29 and 2.32. When it is applied to 1D problem, the reflection and transmission coefficient confined in a region between x_a and x_b can be found. Here, the potential V = 0 at x < x_a or x > x_b and the total wavefunction obtained for the incident and scattered wave by solving schrodinger equation, is,

$$\psi(x) = \psi_i(x) + \psi_s(x), \tag{2.33}$$

here

$$\psi_i(x) = \psi_i exp[ik_0(x - x_a)], \tag{2.34}$$

where

$$k_0 = \sqrt{\frac{2mE}{\hbar^2}}. \tag{2.35}$$

Since the incident wave ($\psi_i(x)$) moves towards the potential barrier from the left, to ensure $\psi_i(x_a) = \psi_i$ an extra factor of $exp(-ik_0x)$ is introduced to $\psi_i(x)$ at x_a. Similarly at point x_b, when the particle attains its shape and the wave is scattered ($\psi_s(x)$), part of its wave is transmitted to the right and partly is reflected back to left such that $\psi_i(x_b) = \psi_i$. The total wavefunction at x_a

and x_b must be proportional to ψ_i, and the wave components will be,

$$\psi(x_a) = \psi_i \begin{bmatrix} 1 \\ r \end{bmatrix}, \tag{2.36}$$

and

$$\psi(x_b) = \psi_i \begin{bmatrix} t \\ 0 \end{bmatrix}. \tag{2.37}$$

Here r and t are reflection and transmission coefficients. To define the reflectance and transmittance moving from left to right in terms of the real quantities is given by,

$$R = \mid r^2 \mid \& T = \mid t^2 \mid, \tag{2.38}$$

and accordingly for the 1D tunneling problem, the transfer matrix relation will be,

$$\begin{bmatrix} t \\ 0 \end{bmatrix} = M(x_b, x_a) \begin{bmatrix} 1 \\ r \end{bmatrix}, \tag{2.39}$$

hence, eqn. 2.39 with regard to transfer matrix components is,

$$r = \frac{M_{21}}{M_{22}}; \quad t = \frac{M_{11}M_{22} - M_{12}M_{21}}{M_{22}} = det(M)/M_{22}. \tag{2.40}$$

2.4 Cubic Plus Proximity model with Improved Transfer Matrix method

In the present work, a well suitable model (CPP-ITM), to study the structure and stability of heavy and superheavy nuclei through alpha and cluster decays is being considered. This model can also be extended to study proton emission. Based on the preliminary examinations done in choosing the best suitable potential for theoretical predictions of SHN, the cubic plus proximity potential seems to be ideal and hence it is coded to study the nuclei in heavy and super heavy regions, $Z \geq 84$. In the study of radioactive decay, alpha or cluster decay of some unstable nuclei can be treated partly like a fission process, where alpha particle exists as an entity in the heavy nucleus, which is in constant motion and is being held inside the nucleus under the influence of nuclear potential. Every time the preformed alpha particle hits the walls of the nucleus there is a higher probability of penetrating through the barrier and giving rise to daughter and fragment nuclei. Proceeding to this study it is assumed that the total potential barrier has two parts: the pre-scission region and post-scission region. In the post-scission region, the total interaction between daughter and fragment nuclei is governed by nuclear interaction, Coulomb interaction, and centrifugal term. Nuclear potential is an important factor in studying nuclear reaction, structure, and decay. Because of its good predictive power the proximity-77 potential may be the best suitable potential. It is mainly composed of two components: (i) the shape and geometry of the two nuclei and (ii) the separation distance between the two nuclei.

To deal with the pre-scission region several authors have adopted different approaches. One way of constructing a suitable pre-scission potential is extending the nuclear potential in the pre-scission region together with the Coulomb potential, and (ii) the other one is the method to use a general n-power law or using analytical expressions for the pre-formation factor (K. P. Santhosh et al., 2015). In the pre-scission region, a cubic polynomial potential was considered by Shanmugam and Kamalakaran (Shanmugam and Kamalaharan, 1988) and which appropriately ensures the conservation of energy to provide good accuracy. For the precise description of splitting dynamics, we employ a well-known form of effective mass in the pre-scission region as the emitted particle and daughter have not attained their individual identity in the pre-scission region and hence using reduced mass in this region is inappropriate. But the effective mass proposed by Moller et al., (Moller et al., 1987) seems to be good alternate.

The proton emission, the alpha decay, and the cluster decay can be explained based on a common mechanism, the tunneling phenomena. In a wide variety of situations, one-dimensional modeling of the tunneling process is adequate and the traditional WKB approximation provides formulas that are easy to use and reasonably accurate but exhibits shortcomings in the regions where the potential changes abruptly and encounters difficulty when the incident energies are close to the barrier height. The model used in this thesis, a recently developed improved transfer matrix method, for evaluating the tunneling probability has considerable improvements over the traditional Transfer matrix (TM) method, where the latter has been employed in the research of semiconductor physics and even nuclear reactions but its use to decay studies has been very limited since it generally involves plane waves. The improved transfer matrix method involving WKB approximation (TM+ WKB)at the boundary of the computational domain has indicated considerable improvement. The test results of parabolic and secant hyperbolic potentials reveal that the improved transfer matrix (TM) are indistinguishable from the exact results, and the relative error is fairly less than the usual plane waves transfer matrix method. Hence, we employed this improved transfer matrix method on tunneling problems due to its accuracy and its different perspective. A detailed discussion on this present model is briefed in the following sections.

2.4.1 Interaction Potential: Cubic plus proximity

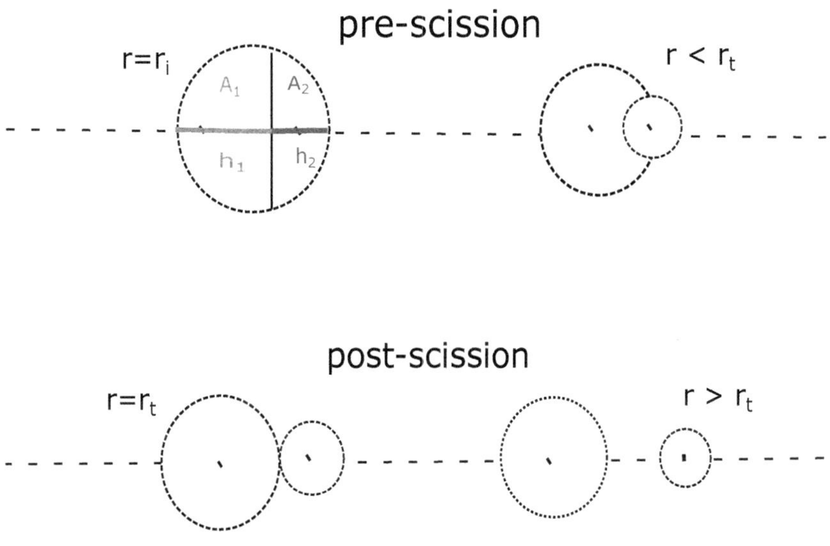

Figure 2.1: Schematic representation of decay process in fission like model.

The decay process involving a light particle or light cluster can be treated as a fission-like process where a parent nuclei gets split into a daughter and fragment nucleus. The process of formation of daughter and fragment is very complex, and this process occurs in the region $r < r_t$ called the pre-scission region where r_t is the touching distance between daughter and fragment and $r > r_t$ is the post-scission region where a fragment has attained its identity and is moving apart from the daughter. Figure 2.1 illustrates the said decay process schematically.

In the post-scission region, the total interaction between daughter and fragment nuclei is governed by nuclear interaction, Coulomb interaction and centrifugal term which is given by

$$V_{ext}(r,\theta,\phi) = V_n(r,\theta,\phi) + V_c(r,\theta,\phi) + V_l(r,\theta,\phi). \tag{2.41}$$

The total interaction potential in the post-scission region at $r = r_t$ and $r > r_t$ between the daughter and fragment nuclei is always a sum of the nuclear part $V_n(r)$ with short-range attractive force, a Coulomb term $V_c(r)$ with long-range repulsive force and a centrifugal part $V_l(r)$, For a spherical nuclei V_{ext} is taken as,

$$V_{ext}(r) = V_n(r) + \frac{Z_1 Z_2 e^2}{r} + \frac{\hbar^2 l(l+1)}{2\mu r^2}, \tag{2.42}$$

where Z_1, Z_2 are the atomic numbers of daughter and fragment nucleus respectively, μ is the reduced mass and l is the angular momentum. Although the Coulomb potential that arises due to the interaction between the charges of daughter $Z_1 e$ and the fragment nuclei $Z_2 e$ and the centrifugal term $V_l(r)$ which is l- dependent and is taken Zero for ground state transition are quite well expressed and calculated accordingly. The nuclear part of potential acting between the two atomic nuclei has seen many formalism over the years. To describe the nuclear interaction between two spherical nuclei, a best suitable proximity potential which closely agrees with the experimental results is taken from Blocki et al., (Błocki et al., 1977) as,

$$V_n(r) = 4\pi\gamma b\bar{R}\phi(\varepsilon). \tag{2.43}$$

Here b (nuclear surface width) is ≈ 1 fm, $\bar{R} = \frac{C_1 C_2}{C_1 + C_2}$ is the reduced radius and $\phi(\varepsilon)$ is the universal proximity potential. $\varepsilon = r - C_1 - C_2/b$ is obtained by measuring separation $(r - C_1 - C_2)$ with nuclear surface width (b). The Sussmann relation (Sussmann, 1973) between the effective radius R_i and central radius C_i can be evaluated as,

$$C_i = R_i\left(1 - \frac{b}{R_i}\right) \text{ fm}. \tag{2.44}$$

To get nuclear sharp radii R_i, a simple semi-empirical relation in terms of mass number A_i is taken as (Błocki et al., 1977),

$$R_i = 1.28A_i^{1/3} - 0.76 + 0.8A_i^{-1/3} \text{ fm } (i = 1, 2). \tag{2.45}$$

In the eqn. 2.43, γ represents the nuclear surface energy or surface tension coefficient, and it can be calculated using Lysekil mass formula (W. D. Myers and Swiatecki, 1966) defined as,

$$\gamma = \gamma_0(1 - k_s I^2) \text{ MeV/fm}^2, \tag{2.46}$$

where $I = \frac{N-Z}{N+Z}$, N and Z refer to the combined system of interacting nuclei. In the above equation, the surface energy constant $\gamma_0 = \frac{a_2}{4\pi r_0^2}$. Here a_2= 17.9439 MeV is the liquid drop model surface energy coefficient, $r_0 \approx 1.2249$ fm is the nuclear radius constant and k_s is the surface asymmetry constant with values $\gamma_0 = 0.9517$ and $k_s = 1.7826$, were obtained by fitting experimental binding energy per particle against the square of relative neutron excess by Myers & Swiatecki (W. D. Myers and Swiatecki, 1966; W. Myers and Swiatecki, 1967), and they have validated the correctness of the said adjustable parameters by studying specifically ^{201}Tl and found to be sufficiently close to the experiment. This parameter set is termed as proximity-77 parametrization.

With the recent literature available, Proximity-77 has exhibited as good predictive power in α-decay half-life calculations among 25 different nuclear potentials (K. P. Santhosh and Sukumaran, 2017a) and for cluster decay 12 different versions of nuclear potentials (K. P. Santhosh and Sukumaran, 2017b). Prox-77 has also been employed for decay studies associated with SHN and provided good agreement with experimental results (K. Santhosh et al., 2009; K. P. Santhosh and Priyanka, 2014). In this thesis we use Proximity-77 nuclear potential to describe the nuclear interaction (Błocki et al., 1977). The universal proximity potential can be written as,

$$\phi(\varepsilon) = -4.41 \exp\left(-\frac{\varepsilon}{0.7176}\right) \quad \text{for} \quad \varepsilon > 1.9475, \tag{2.47}$$

$$\phi(\varepsilon) = -1.7817 + 0.9270\varepsilon + 0.0169\varepsilon^2 - 0.05148\varepsilon^3 \quad \text{for} \quad 0 \leq \varepsilon \leq 1.9475. \tag{2.48}$$

Defining the potential in the pre-scission region is quite complex, and so different authors have followed varied approaches. In this work the cubic polynomial of Shanmugam and Kamalakaran (Shanmugam and Kamalaharan, 1988) is being considered for $r < r_t$ ie., the pre-scission region and the form of potential is taken appropriately to ensure the conservation of energy,

$$V_{ov}(r) = (-E_v + Q) + [V_{ext}(r_t) + E_v - Q] \times \left[s_1 \left(\frac{r - r_i}{r_t - r_i} \right)^2 - s_2 \left(\frac{r - r_i}{r_t - r_i} \right)^3 \right]. \quad (2.49)$$

At r_i the potential is $-E_v + Q$ and at r_t the potential is $V_{ext}(r_t) = V_n(r_t) + V_c(r_t) + V_l(r_t)$. Applying the boundary condition that the value of potential and its first derivative are continuous at r_i and r_t then $s_1 = 3 - s$, $s_2 = 2 - s$. The end-point values are in agreement with ref. (Shanmugam and Kamalaharan, 1988) studied with respect to Q-value as origin.

$$s = \frac{r_t - r_i}{V_{ext}(r_t) + E_v - Q} V'_{ext}(r_t), \quad (2.50)$$

and r_i which is the distance between the center of mass of daughter and fragment within the parent nuclei, which is obtained by the planar section cut method described in ref. (Shanmugam and Kamalaharan, 1988). The parent nucleus is cut into two unequal portions with the masses of the light and heavy nuclei and determined the relationship between the heights h_1 and h_2 of such light and heavy portions which is given as

$$h_1 + h_2 = 2R_0. \quad (2.51)$$

If V_1 and V_2 are the volumes of the fragments with mass numbers A_1 and A_2 respectively and V is the volume of the spheroidal parent nucleus then using the relation $V_1/V = A_1/A$, it can be written as,

$$\frac{3h_1}{R_0} + \left(1 - \frac{h_1}{R_0}\right)^3 - 1 = 4\frac{A_1}{A}, \quad (2.52)$$

where h_1 can be found numerically. Thus the expression for r_i can be given as

$$r_i = \frac{3}{4} \left[\frac{h_1^2}{R_0 + h_1} + \frac{h_2^2}{R_0 + h_2} \right]. \quad (2.53)$$

The lowest possible energy that a quantum mechanical system can have is the zero-point vibrational energy E_v, which is the finite amount of kinetic energy retained by the particle which can exist even at any moment of the decay process. The value of E_v for decay studies is obtained using the eqn. 2.3 (D. N. Poenaru et al., 1985). The parameters of the above relation were obtained by the best fit with a selected set of experimental data of 379 alpha emitters and one cluster radioactive ^{14}C emission from ^{222}Ra, which consist of majorly heavy isotopes and two superheavy isotopes of Z=103 and 105 (D. N. Poenaru et al., 1985). The theoretical half-lives from Analytical Superasymmetric Fission Model (ASAFM) on comparison with experimental half-lives provided the suitable parameters of $\frac{E_v}{Q}$ relation. This relation has been employed in different decay half-life models and the results of α or cluster decay half-lives are seen to hold good for superheavy nuclei (Samanta, 2008; D. Poenaru and Ivascu, 1984; D. Poenaru et al., 1986; D. N. Poenaru et al., 1985). For proton decay $E_v = 0.1045Q_p$ which includes odd-even staggering effect is employed (Guo and Zhang, 2014; D. N. Poenaru et al., 1986).

In the pre-scission region, the dynamics of splitting can be taken into account (in $r < r_t$ the fragment would not have attained its separate identity) by a suitable expression for effective mass given by Moller et al., (Moller et al., 1987)

$$\mu_{eff} = \mu + \frac{17}{15}\mu \exp\left[-\frac{128}{51}\left(\frac{r - r_i}{R_0}\right)\right] \times 16f(r,r_t), \tag{2.54}$$

where

$$f(r,r_t) = \begin{cases} \left[\frac{r_t - r}{r_t - r_i}\right]^4, & \text{if } r \leq r_t \\ 0, & r \geq r_t \end{cases}$$

The effective mass is a decreasing function and at $r = r_t$ the effective mass $\mu_{eff} = \mu$ where $\mu = \frac{A_1A_2}{A_1+A_2}m_n$.

2.4.2 Barrier Penetrability : Improved transfer matrix method

The improved transfer matrix method with WKB approximation at the boundaries is best suited and accurate (Biswas and Kumar, 2014) than the usual transfer matrix method (consisting only the plane waves (Zanganeh and Wang, 2014; Wu et al., 2018)). A short description of this new method is given in this section. The potential barrier $V(r)$ experienced by a particle with mass

$\mu(r)$ can be split into N segments(as illustrated in fig. 2.2) in which potential energy can be taken as constant.

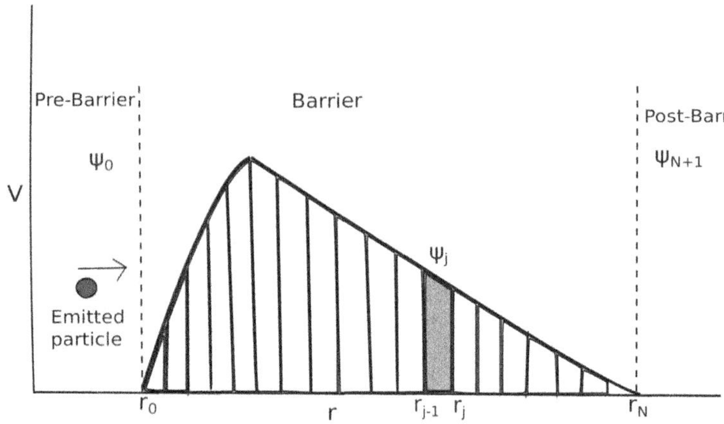

Figure 2.2: Schematic representation of barrier, pre-barrier and post-barrier regions

With multi-step function, one can represent

$$V(r) = V_j = V\left[\frac{r_{j-1}+r_j}{2}\right], \tag{2.55}$$

where $r_{j-1} < r < r_j$ where $j = 1,2,...N$. The wavefunction ψ_j in the barrier region j can be obtained as

$$\psi_j(r) = A_j \exp(ik_j r) + B_j \exp(-ik_j r), \tag{2.56}$$

where potential V_j is constant, and having energy Q and

$$k_j = \frac{1}{\hbar}\sqrt{2\mu(Q-V)}. \tag{2.57}$$

The first order WKB wavefunction in pre-barrier region is given by

$$\Psi_0^{wkb}(r) = \frac{A_0}{\sqrt{\hbar k(r)}} e^{i\int^r k(r')} dr' + \frac{B_0}{\sqrt{\hbar k(r)}} e^{-i\int^r k(r')} dr'. \tag{2.58}$$

At boundary r_0, requiring continuity of ψ and $\frac{d\psi}{dr}$ we get

$$\begin{bmatrix} A_1 \\ B_1 \end{bmatrix} = M_0 \begin{bmatrix} A_0 \\ B_0 \end{bmatrix}, \tag{2.59}$$

where

$$M_0 = \frac{1}{2\sqrt{\hbar k_0}} \begin{bmatrix} \left(1+S_0^+\right) e^{-ik_1 r_0} & \left(1-S_0^-\right) e^{-ik_1 r_0} \\ \left(1-S_0^+\right) e^{ik_1 r_0} & \left(1+S_0^-\right) e^{ik_1 r_0} \end{bmatrix}, \tag{2.60}$$

and

$$S_0^{\pm} = \frac{k_0}{k_1} \mp \frac{ik_0'}{k_0 k_1}, \tag{2.61}$$

with $k_0' = -\mu V'(x_0)/k_0\hbar^2$. Inside the barrier region, A_j and B_j can be expressed as multiplication of matrices, as

$$\begin{bmatrix} A_j \\ B_j \end{bmatrix} = \prod_{y=0}^{j-1} M_y \begin{bmatrix} A_0 \\ B_0 \end{bmatrix}, \tag{2.62}$$

$$M_y = \frac{1}{2} \begin{bmatrix} \left(1+S_y\right) e^{-i(k_{y+1}-k_y)r_y} & \left(1-S_y\right) e^{-i(k_{y+1}+k_y)r_y} \\ \left(1-S_y\right) e^{i(k_{y+1}+k_y)r_y} & \left(1+S_y\right) e^{i(k_{y+1}-k_y)r_y} \end{bmatrix}, \tag{2.63}$$

and

$$S_y = \frac{k_y}{k_{y+1}} \text{for y=1,2,...N-1.} \tag{2.64}$$

The first order WKB wavefunction in post-barrier region is given by

$$\Psi_{N+1}^{wkb}(r) = \frac{A_{N+1}}{\sqrt{\hbar k(r)}} e^{i\int^r k(r')} dr' + \frac{B_{N+1}}{\sqrt{\hbar k(r)}} e^{-i\int^r k(r')} dr'. \tag{2.65}$$

At boundary r_N requiring continuity of ψ and $\frac{d\psi}{dr}$ we get

$$\begin{bmatrix} A_{N+1} \\ B_{N+1} \end{bmatrix} = M_N \begin{bmatrix} A_N \\ B_N \end{bmatrix}, \tag{2.66}$$

$$M_N = c_N \begin{bmatrix} \left(iS_N^+ + G_N\right) e^{ik_N r_N} & \left(iS_N^- + G_N\right) e^{-ik_N r_N} \\ \left(iS_N^+ - G_N\right) e^{ik_N r_N} & \left(iS_N^+ - G_N\right) e^{-ik_N r_N} \end{bmatrix}, \tag{2.67}$$

and

$$S_N^{\pm} = k_{N+1} \pm k_N, G_N = k'_{N+1}/2k_{N+1}, \tag{2.68}$$

where $k'_{N+1} = -\mu V'(r_N)/k_{N+1}\hbar^2$ and $c_N = \frac{\sqrt{\hbar}}{2i\sqrt{K_{N+1}}}$. By fixing $A_0 = 1$ and $B_{N+1} = 0$ for $j = N+1$ the transmission amplitude is given by

$$A_{N+1} = \frac{k_1}{k_N} \frac{det(M_0)det(M_N)}{M_{22}}, \tag{2.69}$$

where M_{22} is matrix element obtained from the result of matrix multiplication, and can be obtained using

$$M = \begin{bmatrix} M_{11} & M_{12} \\ M_{21} & M_{22} \end{bmatrix} = \prod_{y=0}^{N} M_y. \tag{2.70}$$

The tunneling probability is given by

$$P = |A_{N+1}|^2. \tag{2.71}$$

The decay constant is given as $\lambda = P\nu$ where ν is the frequency of assaults on the barrier per second. From the classical description $\nu = \frac{1}{2R}\sqrt{\frac{2Q}{\mu}}$. The total tunnelling probability through the barrier can be split as $P = P_{ov}P_{ext}$. The P_{ov} can be interpreted as preformation probability which is the probability of fragment being preformed within the parent (D. N. Poenaru and Greiner, 1991). The half-life of nuclei which decay through particle emission or cluster emission is calculated using eqn. 2.4. To establish a confident over the execution power of our improved transfer matrix method, the results obtained using experimental Q-values for the alpha decay half-lives of superheavy isotopes are compared with the experimental Q values and WKB method and are tabulated in table 2.2. It is evident that, though the order of tunnelling probability calculated with the WKB method and ITM are same, the $T_{1/2}$ values seems to be different in WKB calculations and the ITM provides very close value to the experiment. In predicting the half life for the said nuclei, for eg., the alpha decay of superheavy isotopes shows the P^{ITM} value as roughly 0.9 - 2.0 times of P^{WKB}. Such improvement in $T_{1/2}$ calculations leads us to proceed for the entire heavy mass region of the periodic table and is presented in this thesis. The calculation done by ITM method is been compared with the empirical relation and formalism used by different authors so that an overall picture about the credibility of the used model, the ITM with the WKB approximation, can be addressed. The empirical relation & formalism used in this thesis for comparison are briefed below.

Table 2.2: Tunneling probability and $T_{1/2}$ by Improved transfer matrix method and WKB approximation.

Parent Nuclei	Q (MeV)	Tunnelling probability P		$logT_{1/2}$ ($T_{1/2}$ in s)		
	Exp	P^{WKB}	P^{ITM}	Exp	WKB	ITM
$^{294}118$	11.82	5.508×10^{-19}	6.967×10^{-19}	-3.16	-3.09	-3.19
$^{293}116$	10.71	3.482×10^{-21}	7.256×10^{-21}	-1.24	-0.87	-1.19
$^{292}116$	10.78	5.241×10^{-21}	9.918×10^{-21}	-1.89	-1.05	-1.33
$^{290}116$	11.00	1.922×10^{-20}	3.652×10^{-20}	-2.08	-1.62	-1.90
$^{287}115$	10.76	8.870×10^{-21}	1.561×10^{-20}	-1.43	-1.28	-1.53
$^{289}114$	9.98	1.272×10^{-22}	2.409×10^{-22}	0.28	0.58	0.30
$^{285}112$	9.32	5.392×10^{-24}	1.137×10^{-23}	1.45	1.97	1.64
$^{283}112$	9.66	6.235×10^{-23}	1.246×10^{-22}	0.62	0.89	0.59
$^{281}112$	10.46	1.560×10^{-20}	3.052×10^{-20}	-1.00	-1.460	-1.75
$^{282}112$	9.16	1.399×10^{-24}	1.606×10^{-24}	2.00	2.55	2.49
$^{279}111$	10.53	4.585×10^{-20}	3.314×10^{-20}	-1.05	-1.99	-1.85
$^{275}108$	9.45	3.857×10^{-22}	5.391×10^{-22}	-0.70	0.10	-0.04

2.4.3 Empirical relation : proton emission

Ni et al., (Ni and Ren, 2012), provided an empirical formula to evaluate the proton emission half-lives independent of l value as

$$LogT_{1/2} = aZ_1Z_2\sqrt{\frac{\mu'}{Q}} + b\sqrt{\mu' Z_1Z_2} + c + d|\beta_2|^3, \tag{2.72}$$

where

$$\mu' = \frac{A_1 A_2}{A_1 + A_2}, \tag{2.73}$$

Thus calculated $T_{1/2}$ suits well with our ground state transition studies and β_2 is taken as zero since we dealt only spherical nuclei for proton emission.

2.4.4 Empirical relation : alpha and cluster emission

Universal decay law (UDL)

A linear decay formula from the microscopic to macroscopic mechanism of charged particle emission was given by Qi et al., (Qi, Xu, Liotta and Wyss, 2009)

$$log_{10}T_{1/2} = aX' + b\rho' + c, \tag{2.74}$$

where

$$X' = Z_1 Z_2 \sqrt{\frac{\mu'}{Q}} \tag{2.75}$$

$$\rho' = \sqrt{\mu' Z_1 Z_2 (A_1^{1/3} + A_2^{1/3})}. \tag{2.76}$$

This equation is obtained from the concepts of Geiger and Nuttal law. μ' is obtained from eqn. 2.73. Here the coefficients $a = 0.4314$, $b = -0.4087$ & $c = -25.7725$ are the constants obtained by fitting the experimental values of α and cluster decays (Qi, Xu, Liotta, Wyss, Zhang et al., 2009). Half-lives are related to the energy released in the decay process. The dependence of Q-value, is the illustration of quantum tunnelling of the alpha cluster through the coulomb barrier.

Viola-Seaborg semi-empirical formula (VSS)

In 1911, Geiger and Nuttal exhibited straight-line fit by plotting the $log_{10}T_{1/2}$ with $Q^{-1/2}$ decay energy. The Viola Seaborg, by generalizing the Geiger and Nuttal law, obtained a relation for half-life by analyzing the experimental half-lives of alpha, cluster and spontaneous fission of heavy elements of mass range A $\geq$ 140 is given by,

$$log_{10}(T_{1/2}) = (aZ + b)Q^{-1/2} + cZ + d + h_{log}. \tag{2.77}$$

here the parameters a, b, c and d obtained from the linear square fit for even-even nuclei and h_{log} is the hindrance factor for unpaired nucleons (Viola and Seaborg, 1966; Sobiczewski et al., 1989). The constants $a = 1.64062$, $b = -8.54399$, $c = -0.19430$, $d = -33.9054$, and the value of h_{log} is taken from (Dong and Ren, 2005).

$$
\begin{aligned}
h_{log} &= 0 && \text{for Z, N= even,}\\
h_{log} &= 0.8937 && \text{for Z = even, N= odd,}\\
h_{log} &= 0.5720 && \text{for Z = odd, N= even,}\\
h_{log} &= 0.9380 && \text{for Z, N = odd.}
\end{aligned}
$$

Royer's analytical formula

The simple relation which connects α-decay half-life and Q-value devised by Geiger and Nuttal was further improved by different authors (Viola and Seaborg, 1966; Sobiczewski et al., 1989; Brown, 1992). One of the most commonly used formulae for calculating the alpha decay half-life is the analytical formula of Royer, (Royer, 2000)

$$
log_{10}T_{1/2} = a + bA^{1/6}\sqrt{Z} + c\frac{Z}{\sqrt{Q}}. \tag{2.78}
$$

The parameters a, b & c were extracted by fitting the experimental data. Fitting with the experimental Q-values of 131 even-even, 106 even-odd, 86 odd-even, and 50 odd-odd nuclei the parameter values obtained were, (-25.31, -1.1629, 1.5864); (-26.65, -1.0859, 1.5848); (-25.68, -1.1423, 1.592) and (-29.48, -1.113, 1.6971) respectively with the RMS deviation values 0.285, 0.39, 0.36 and 0.35 respectively.

Horoi's empirical formula

For cluster decay, a suitable model-dependent description containing dynamics of the decay process was proposed by Horoi (Horoi, 2004; K. P. Santhosh et al., 2008). The decay half-life has linear dependents on the square root of the reduced mass of daughter-cluster combination and the scaling factor. The empirical formula takes the form,

$$logT_{1/2} = (a_1\mu'^x + b_1)((Z_1Z_2)^y\frac{1}{\sqrt{Q}} - 7) + (a_2\mu'^x + b_2) \tag{2.79}$$

By fitting with experimental data, the constants reported are $a_1 = 9.1$, $b_1 = -10.2$, $a_2 = 7.39$, $b_2 = -23.2$, $x = 0.416$, $y = 0.613$.

2.4.5 Empirical relation: spontaneous fission

The two main decay modes of superheavy nuclei are the alpha particle emission succeeded by spontaneous fission. Spontaneous fission shares the same mechanism of quantum tunnelling as of alpha decay. In comparison to the alpha decay, spontaneous fission is quite complicated with large uncertainties in the fission process such as the emitted energy, number of neutron being emitted and the charge and mass numbers of the emitted fragments. For the SHN undergoing spontaneous fission, $T_{1/2}$ can be calculated using a generalized liquid drop model. The semi-empirical formula by Xu et al., (Xu et al., 2008; Xu and Ren, 2005) for even-even nuclei, is as follows.

$$log_{10}T_{1/2} = \exp[2\pi(c_0 + c_1A + c_2Z^2 + c_3Z^4 + c_4(N-Z)^2 - (0.13323Z^2A^{-1/3} - 11.64))]$$

By fitting with experimental data the values of coefficient, reported by Xu et al., for even-even nuclei are $c_0 = -195.09227$, $c_1 = 3.10156$, $c_2 = -0.04386$, $c_3 = 1.4030 \times 10^{-6}$, and $c_4 = -0.03199$ (Xu et al., 2008; Xu and Ren, 2005).

The above given half-life formulae are been used throughout the thesis wherever SHN are dealt with.

2.5 Summary

To summarise, this chapter focuses on essential elements, the potential and the tunneling probability in order to construct a reliable theoretical model for heavy nuclei and SHN decay investigations, and for which the CPP-ITM approach is established with $V(r) = V_n^{Prox77}(r) + V_c(r) + V_l(r)$, $r \geq r_t$; $V(r) = V_{ov}^{Cubic}(r)$, $r < r_t$, for its accuracy and different perspective on decay study. This fission-like model is employed for estimating half-lives alongside empirical formulas for proton emission (Ni), alpha decay (UDL, VSS, Royer), and cluster decay (UDL, Horoi).

References

Aleksandrov, D., Belyatskij, A., Glukhov, Y. A., Nkol'skij, E. Y., Novatskij, B., Ogloblin, A. and Stepanov, D. (1984). Observation of spontaneous emission of ^{14}C nuclei from ^{223}Ra. *Pis' ma v Zhurnal Ehksperimental'noj i Teoreticheskoj Fiziki*, **40**, 152–154.

Alsaif, N. A., Saleh Ahmed, S. M. and Al-Ghamdi, H. (2020). Alpha clustering preformation probability in even-even and odd-A 270317(116 and 117) using cluster formation model and the mass formulae : KTUY05 and WS4. *Results in Physics*, **19**, 103689.

Biswas, D. and Kumar, V. (2014). Improved transfer matrix methods for calculating quantum transmission coefficient. *Physics Review E*, **90**, 013301.

Blendowske, R., Fliessbach, T. and Walliser, H. (1987). Microscopic calculation of the ^{14}c decay of ra nuclei. *Nuclear Physics A*, **464**, 75–89.

Błocki, J., Randrup, J., Świątecki, W. J. and Tsang, C. F. (1977). Proximity forces. *Annals of Physics*, **105**, 427-462.

Brown, B. A. (1992). Simple relation for alpha decay half-lives. *Physical Review C*, **46**, 811-814.

Buck, B., Merchant, A. and Perez, S. (1992). α decay calculations with a realistic potential. *Physical Review C*, **45**, 2247.

Deng, J.-G. and Zhang, H.-F. (2021). Correlation between α-particle preformation factor and α decay energy. *Physics Letters B*, **816**, 136247.

Dong, T. and Ren, Z. (2005). New calculations of α-decay half-lives by the Viola-Seaborg formula. *The European Physical Journal*, **26**, 69-72.

Gales, S., Hourani, E., Hussonnois, M., Schapira, J., Stab, L. and Vergnes, M. (1984). Exotic nuclear decay of ^{223}Ra by emission of ^{14}C nuclei. *Physical Review Letters*, **53**, 759.

Guo, C. L. and Zhang, G. L. (2014). Analysis of proton radioactivity of nuclei by using proximity potential with a new universal function. *The European Physical Journal A*, **50**, 187.

Gupta, R. K. and Greiner, W. (1994). Cluster radioactivity. *International Journal of Modern Physics E*, **3**, 335–433.

Horoi, M. (2004). Scaling behaviour in cluster decay. *Journal of Physics G: Nuclear and Particle Physics*, **30**, 945–955.

Kramers, H. A. (1926). Wellenmechanik und halbzahlige quantisierung. *Zeitschrift für Physik*, **39**, 828–840.

Moller, P., Nix, J. R. and Swiatecki, W. J. (1987). Calculated fission properties of the heaviest elements. *Nuclear Physics A*, **469**, 1.

Myers, W. and Swiatecki, W. (1967). Anomalies in nuclear masses. *Arkiv för Fysik*, **36**, 343–352.

Myers, W. D. and Swiatecki, W. J. (1966). Nuclear masses and deformations. *Nuclear Physics A*, **81**, 1 - 60.

Ni, D. and Ren, Z. (2012). New formula of half-lives for proton emission from spherical and deformed nuclei. *Romanian Journal of Physics*, **57**, 407-418.

Nix, J. R. (1969). Further studies in the liquid-drop theory on nuclear fission. *Nuclear Physics A*, **130**, 241–292.

Poenaru, D., Greiner, W., Depta, K., Ivascu, M., Mazilu, D. and Sandulescu, A. (1986). Calculated half-lives and kinetic energies for spontaneous emission of heavy ions from nuclei. *Atomic Data and Nuclear Data Tables*, **34**, 423 - 538.

Poenaru, D. and Ivascu, M. (1984). *New nuclear decay modes by spontaneous and beta-enhanced emission of heavy ions* (Tech. Rep.). Institutul Central de Fizica.

Poenaru, D., Ivascu, M. and Sandulescu, A. (1979). Alpha decay as a fission-like process. *Journal of Physics G: Nuclear Physics*, **5**, L169.

Poenaru, D. N. and Greiner, W. (1991). Rare decay modes by cluster emission from nuclei. *Journal of Physics G: Nuclear and Particle Physics*, **17**, S443.

Poenaru, D. N., Greiner, W., Ivaşcu, M., Mazilu, D. and Plonski, I. H. (1986). Odd-even staggering of heavy cluster spontaneous emission rates. *Zeitschrift für Physik A Atomic Nuclei*, **325**, 435.

Poenaru, D. N., Ivaşcu, M., Sndulescu, A. and Greiner, W. (1985). Atomic nuclei decay modes by spontaneous emission of heavy ions. *Physical Review C*, **32**, 572–581.

Price, P., Stevenson, J., Barwick, S. and Ravn, H. (1985). Discovery of radioactive decay of ^{222}Ra and ^{224}Ra by ^{14}C emission. *Physical review letters*, **54**, 297.

Qi, C., Xu, F. R., Liotta, R. J. and Wyss, R. (2009). Universal decay law in charged-particle emission and exotic cluster radioactivity. *Physical Review Letters.*, **103**, 072501.

Qi, C., Xu, F. R., Liotta, R. J., Wyss, R., Zhang, M. Y., Asawatangtrakuldee, C. and Hu, D. (2009). Microscopic mechanism of charged-particle radioactivity and generalization of the Geiger-Nuttall law. *Physical Review C*, **80**, 044326.

Rose, H. J. and Jones, G. A. (1984). A new kind of natural radioactivity. *Nature*, **307**, 245-247.

Royer, G. (2000). Alpha emission and spontaneous fission through quasi-molecular shapes. *Journal of Physics G: Nuclear and Particle Physics*, **26**, 1149–1170.

Samanta, C. (2008). Superheavy elements in the magic islands. In *Fission and properties of neutron-rich nuclei* (pp. 329–337). World Scientific.

Săndulescu, A., Gupta, R. K., Greiner, W., Carstoiu, F. and Horoi, M. (1992). Role of cluster deformations on their preformation probabilties in radioactive cluster decay studies. *International Journal of Modern Physics E*, **1**, 379–388.

Santhosh, K., Sahadevan, S. and Biju, R. (2009). Alpha radioactivity in heavy and super heavy elements. *Nuclear Physics A*, **825**, 159-174.

Santhosh, K. P., Biju, R. K. and Joseph, A. (2008). A semi-empirical model for alpha and cluster radioactivity. *Journal of Physics G: Nuclear and Particle Physics*, **35**, 085102.

Santhosh, K. P. and Jose, T. A. (2021). Cluster pre-formation probabilities and decay half-lives for trans-lead nuclei using modified generalised liquid drop model (mgldm). *Pramana – Journal of Physics*, **95**, 1–17.

Santhosh, K. P. and Priyanka, B. (2014). α decay chains from Z = 118 superheavy nuclei in the range $271 \leq A \leq 310$. *Physical Review C*, **89**, 064604.

Santhosh, K. P. and Sukumaran, I. (2017a). Alpha decay studies on po isotopes using different versions of nuclear potentials. *The European Physical Journal A*, **53**, 246.

Santhosh, K. P. and Sukumaran, I. (2017b). Studies on cluster decay from trans-lead nuclei using different versions of nuclear potentials. *The European Physical Journal A*, **53**, 136.

Santhosh, K. P., Sukumaran, I. and Priyanka, B. (2015). Theoretical studies on the alpha decay of $^{178-220}$Pb isotopes. *Nuclear Physics A*, **935**, 28-42.

Shanmugam, G. and Kamalaharan, B. (1988). Application of a cubic barrier in exotic decay studies. *Physical Review C*, **38**, 1377.

Shi, Y.-J. and Swiatecki, W. (1985a). Estimates of radioactive decay by the emission of nuclei heavier than α-particles. *Nuclear Physics A*, **438**, 450–460.

Shi, Y.-J. and Swiatecki, W. (1985b). Theoretical estimates of the rates of radioactive decay of radium isotopes by ^{14}C emission. *Physical Review Letters*, **54**, 300.

Sobiczewski, A., Patyk, Z. and Ćwiok, S. (1989). Deformed superheavy nuclei. *Physics Letters B*, **224**, 1 - 4.

Sussmann, G. (1973). Nuclear physics. *Lawrence Berkeley Laboratory University of California Berkeley, California*, **9472**.

Viola, V. and Seaborg, G. (1966). Nuclear systematics of the heavy elements—ii lifetimes for alpha, beta and spontaneous fission decay. *Journal of Inorganic Nuclear Chemistry*, **28**, 741 - 761.

Wu, S., Qian, Y. and Ren, Z. (2018). Half-lives of α-decaying nuclei in the medium-mass region within the transfer matrix method. *Physical Review C*, **97**, 054316.

Xu, C. and Ren, Z. (2005). Systematical law of spontaneous fission half-lives of heavy nuclei. *Physical Review C*, **71**, 014309.

Xu, C., Ren, Z. and Guo, Y. (2008). Competition between α decay and spontaneous fission for heavy and superheavy nuclei. *Physical Review C*, **78**, 044329.

Zanganeh, V. and Wang, N. (2014). Temperature-dependent potential in alpha-decay process. *Nuclear Physics A*, **929**, 94 - 101.

Zhang, H. F., Dong, J. M., Royer, G., Zuo, W. and Li, J. Q. (2009). Preformation of clusters in heavy nuclei and cluster radioactivity. *Physical Review C*, **80**, 037307.

CHAPTER 3

PARTICLE AND LIGHT CLUSTER DECAY OF SUPERHEAVY NUCLEI

3.1 Introduction

Superheavy elements also called transactinide elements beyond $Z > 103$, are produced in laboratories by emitting alpha particles in a series of decay that end in spontaneous fission (SF). Alpha and light cluster decay being important decay modes, which play a pivotal role in decay studies. Investigations on the decay properties of superheavy nuclei (SHN) unveil the degree of stability, as well as the probable decay modes, which shed light on the cluster and spontaneous emission. Also it aid physicists in their investigations of magic numbers, the occurrence of islands of stability, and nuclear structure studies. The observations of two main decay modes-α-decay and SF, provide a good quality of knowledge about SHN's basic properties. The unique signature associated with the identification of new nuclei to be synthesized, is from the α-decay chain of the decaying nuclei. The alpha decay chain provides us the information about the nuclear stability and the structure of not only the originating parent nuclei but also the subsequent parents which are being formed along the chain.

It is worth noting that alpha decay is not just the mode of decay in heavy and super heavy nuclei; literature are abundant on proton emission, cluster decay & SF and numerous microscopic and phenomenological methodologies have been developed to investigate these decay modes. A common mechanism of tunneling events can be used to explain proton emission, alpha decay, and cluster decay. In recent years, extensive investigations on the decay studies of SHN have been performed using the said methodologies (Sharma and Kaur, 2015) (Santhosh and Priyanka, 2014), (Y. Zhang and Wang, 2017), (Poenaru and Gherghescu, 2018; Carmel Vigila Bai and Agnes, 2017). The newly synthesized SHN which usually ends with SF is more complicated as compared to other decay modes. Extensive studies on SF using various theoretical approaches since its discovery (Poenaru and Gherghescu, 2014, 2016, 2017) helped Xu et al., (Xu et al.,

2008; Xu and Ren, 2005) develop a reliable empirical formula for finding the half-life using the parabolic potential approximation, taking nuclear structure effects into account. Also, in superheavy region cluster radioactivity studies have drawn more attention than alpha decay. Cluster decay being an intermediate state between the alpha decay and SF it was first theoretically forwarded by Sandulescu et al., in 1980 (Sandulescu et al., 1980) and its first experimental observation was made by Rose and Jones (Rose and Jones, 1984). Other theoretically predicted lifetime calculations on SHN foresee the cluster emission to dominate over alpha decay (Poenaru and Gherghescu, 2018) and being stronger. SHN can emit proton if the nuclei are present well below the proton drip line. The first radioactive proton emission was observed from ^{53}Co in 1970 (Jackson et al., 1970). Many theoretical works have been framed to understand the emission of protons from heavy and SHN (Santhosh and Sukumaran, 2018; Guo and Zhang, 2014). Also, studies on proton emission have served the understanding of nuclear deformation, masses, and structure. Taking inspiration from previous findings by various authors that focused specifically on heavy & SHN, this chapter has aimed to study the alpha decay & cluster radioactivity via asymmetric fission process, probable proton emissions, and its decay properties using the Cubic plus proximity potential model with Improved transfer matrix method. Part one of this chapter focuses on the decay modes of SHN, while later sections of the chapter examine α-decay chains in the superheavy area. The major goal of this chapter is to propose the existence of proton & neutron shell closure and predict dominant decay modes in the SHN region, which could have an impact on future studies aimed at identifying heavier isotopes to be synthesized in laboratories.

SHN are identified with their α-decay chains. Santhosh and Nithya studied the α-decay chains for several isotopes up to $Z = 132$ and a limited number of isotopes in the range $Z = 132 - 136$ using Coulomb plus proximity model for deformed nuclei(CPPMDN) (Santhosh and Nithya, 2018b,a). We present the calculations on the decay properties of 368 isotopes of nuclei far from the beta stability line and beyond the islands of stability (Z = 130-144) and are studied within the range $290 \leq A \leq 380$, wherein 1-neutron, 2-neutron, 1-proton, 2-proton, alpha particle, the very light cluster of alpha multiple and SF have been considered using a Cubic Plus Proximity model with an improved transfer-matrix method for the first time. This is described in section 3.2. In section 3.3 we present our calculations on the alpha decay chain originating from yet-to-be-discovered SHE Z = 126 and Z = 138, which are theoretically predicted as proton shell closures appearing in different mass models. Also, alpha decay chain study on isotopes of Z = 122 which will be likely to be synthesized in near future is presented. This prediction may pave a way to detecting/synthesizing the most probable isotopes of the SHE Z = 122, 126 & 138 in the laboratory through future experiments. In section 3.4 a brief analysis of decay modes of hyper-heavy nuclei Z = 164 is presented.

3.2 Decay properties of superheavy nuclei Z=130-144

In this work, the decay modes of $Z = 130$ to $Z = 144$ are investigated using the improved transfer matrix approach (Biswas and Kumar, 2014). The potential barrier is divided into multiple small sections in this method, and the wavefunction is assessed in each of them. In all of these regions, the matrix is built using the wavefunction's continuity and derivative. Finally, the tunneling amplitude and probability are calculated using the wavefunctions of the pre-barrier and post-barrier areas. To demonstrate such a problem a proper selection of interaction potential is essential since it must have a good correlation with nuclear force properties observed in studies. Theoreticians have employed several different types of interaction potentials, to discover the optimal nuclear potential suitable alpha and cluster decay (Santhosh and Sukumaran, 2017a,b). To characterize nuclear interaction in this study, we utilize the Proximity-77 incorporated in the Cubic Plus Proximity Potential model discussed in chapter 2 for the calculation.

Ground state properties of nuclei

a) Q-value

The nuclear masses being one of the fundamental quantities in nuclear physics, are important not only for understanding nuclear structure and reactions, but also for understanding the genesis of elements in the universe. Nuclear mass formulas are essential for determining global nuclear parameters as well as investigating exotic structures of exceptionally neutron-rich nuclei, such as SHN and their decay characteristics (Wang et al., 2014).

The binding energy of nuclei is an important quantity associated with a nucleus. Using liquid drop model Bethe-Weizsacker obtained a semi-empirical formula to find the binding energy of nucleus wherein contributions from volume energy, surface energy, Coulomb energy, asymmetry energy, and pairing energy terms were summed up (Krane, 1988). To describe the nuclear mass and binding energy across the nuclear chart, different approaches were adopted to get good estimates for binding energy. The simple approach found in the literature is to modify the Bethe-Weizsacker formula to accommodate shell closure effects (Dong and Ren, 2005, 2008; Samanta and Adhikari, 2002). Moller et al., (Moller et al., 2016) developed finite-range liquid drop model (FRDM) calculations based on the finite-range droplet macroscopic and the folded-Yukawa single-particle microscopic nuclear-structure models over the past two decades; binding energy data up to $Z = 136$ based on FRDM is available in ref. (Moller et al., 2016). Inspired by the Skyrme energy-density functional a macroscopic-microscopic Weizsacker-Skyrme (WS) mass formula was given by Wang et al., (Wang et al., 2014); the binding energy data up to $Z = 132$ is available in Ref. (Wang et al., 2014). Beyond $Z \geq 136$ binding energy data aren't available, to avoid such complications, mean field models have proved to be quite successful in describing the nuclear matter present, spanning from atomic nuclei to neutron stars. To find Q_α, a simple mathematical formula is given as,

$$Q_\alpha(A,Z) = B(A,Z) - B(A-2,Z-2) - B(^4_2He) \tag{3.80}$$

To start with the present study, the DIRHB computer code (Nikšić et al., 2014) which is a relativistic self-consistent mean-field framework code that solves the stationary relativistic Hartree-Bogoliubov equations is taken. A preliminary investigation using DDPC1 effective interaction (Lalazissis et al., 2005) was made to find best suitable Q-value and for it $Z = 118-120$ with neutron number of parent nuclei ranging from $N = 130-250$ are calculated using different mass models WS4 , FRDM , GHIB, KTUY (Cui et al., 2018) and DIRHBS-DD-PC1 (using Dirac Hartree-Bogoliubov mass formula) are listed in table 3.1. For $Z = 118$ it is observed that the Q values of DIRHBS (DD-PC1) with respect to WS4 and FRDM are closely comparative when plotted against the parent nuclei neutron number fig. (3.1). It is evident from fig. 3.1 that the dips seen corresponds to the magic number of the parent nuclei.

Table 3.1: Comparison of Q-values from different mass models, WS4, FRDM, KTUTY and GHFB (Cui et al., 2018) with DD-PC1 (Nikšić et al., 2014).

Z	A	Q-value (MeV)				
		WS4	FRDM	KTUY	GHFB	DIRHBS (DD-PC1)
118	289	12.59	12.76	11.79	12.32	11.33
118	290	12.6	12.67	11.65	12.22	11.20
118	291	12.42	12.55	11.6	10.9	11.65
118	292	12.24	12.39	11.47	10.58	12.11
118	293	12.24	12.34	11.28	11.44	12.13
118	295	11.9	11.93	11.11	10.23	12.07
118	296	11.75	12.28	10.95	10.15	12.03
118	297	12.1	12.39	11.02	11.01	12.00
118	298	12.18	12.49	11.12	10.85	11.97
118	299	12.05	12.51	11.04	10.08	11.96
118	300	11.96	12.51	11.04	10.17	11.98
118	301	12.02	12.56	11.03	10.06	12.05
118	302	12.04	12.62	10.92	10.02	12.29
118	303	12.6	13.38	11.7	11.84	12.71
118	304	13.12	13.4	12.44	12.32	12.92
119	290	13.07	13.31	12.31	12.92	11.76
119	291	13.05	13.24	12.17	12.78	11.62
119	292	12.9	13.08	12.1	11.33	12.13
119	293	12.72	12.92	11.99	10.99	12.66
119	294	12.73	12.85	11.85	11.88	12.68
119	295	12.76	12.94	11.71	11.21	12.66
119	296	12.48	12.98	11.51	10.62	12.62
119	297	12.42	12.9	11.29	10.5	12.59
119	298	12.71	13.09	11.33	11.41	12.55
119	299	12.76	13.08	11.48	11.24	12.53
119	300	12.57	13.03	11.4	10.44	12.52
119	301	12.43	13.08	11.35	10.51	12.54
119	302	12.43	13.05	11.31	10.6	12.62
119	303	12.42	13.11	11.22	10.53	12.87
119	304	12.93	13.86	12.03	12.27	13.27
119	305	13.42	13.86	12.82	12.72	13.44
120	291	13.51	13.87	12.82	12.53	12.20
120	292	13.47	13.78	12.72	12.35	12.05
120	293	13.4	13.65	12.57	10.85	12.63
120	294	13.24	13.49	12.5	10.57	13.23
120	295	13.27	13.46	12.36	11.42	13.25
120	296	13.34	13.59	12.23	11.36	13.23
120	297	13.14	13.65	11.98	11.06	13.21
120	298	13.01	13.24	11.63	11.26	13.18
120	299	13.26	13.73	11.67	11.89	13.16
120	300	13.32	13.7	11.89	11.72	13.15
120	301	13.06	13.62	11.84	10.88	13.15
120	302	12.89	13.55	11.8	11.12	13.17
120	303	12.81	13.51	11.69	10.96	13.26
120	304	12.76	13.55	11.52	10.88	13.50
120	305	13.28	14.26	12.39	12.74	13.87
120	306	13.79	14.28	13.23	13.18	14.02

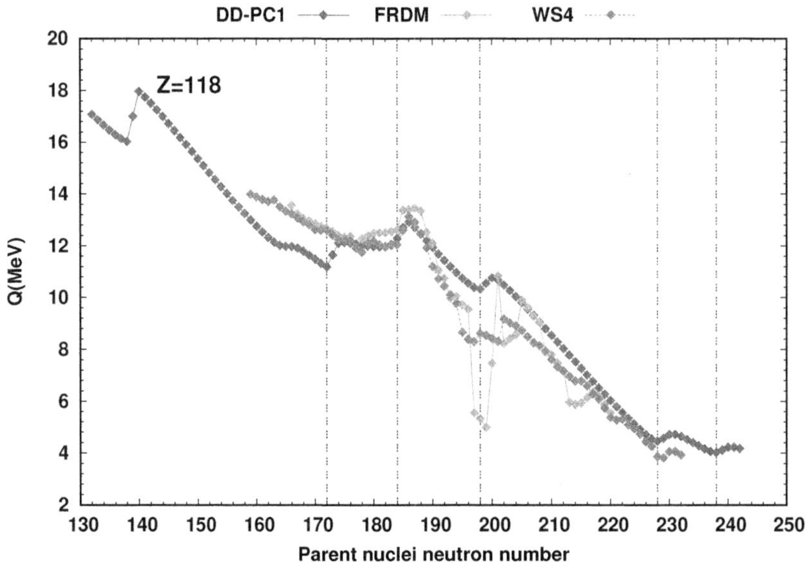

Figure 3.1: Comparison of alpha decay Q-values for $Z = 118$ isotopes obtained from DD-PC1 parametrization, FRDM and WS4 mass models.

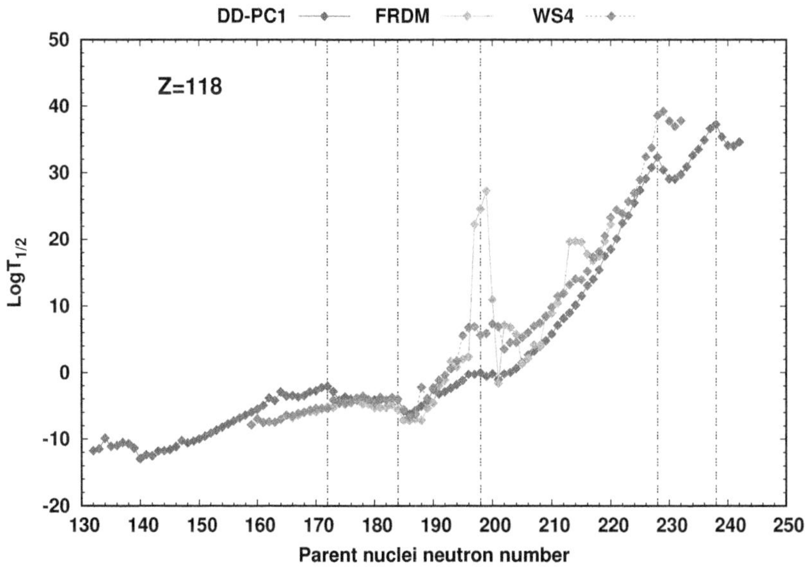

Figure 3.2: $\log T_{1/2}$ value for $Z = 118$ isotopes evaluated with present model of Q-values deduced from DD-PC1 parametrization, FRDM and WS4 mass models.

The magic number at N=184 and 228 is common for WS4 and DIRHBS, at $N = 178$, WS4 and FRDM predicts its magic number nuclei but DIRHBS indicates at N=172. Similarly WS4 shows its dip at N=196 and DIRHBS and FRDM at N=198. Such a similar trend is seen from fig. 3.2 when the half-lives ($T_{1/2}$) of alpha decay is plotted against the parent nuclei neutron number with peaks at $N = 172, 184, 198, 228$ and 238.

b) Particle separation Energy

Nucleon separation energy is the smallest amount of energy required to remove a proton or neutron from a nucleus. It is similar to atomic ionisation energy, but for nuclei. They are also crucial in the investigation of nuclear shell structures. The 1p, 1n, 2p, and 2n separation energies are calculated using eqn. 2.9. Separation energy can be related to Q-value for a given nuclear reaction process from the amount of energy it releases being positive or negative and thus one can estimate its feasibility. The Q-value for a nucleon is the negative of separation energy. Based on the outcome to the preliminary studies, the decay properties for $Z = 130 - 144$ within $290 \leq A \leq 380$ considered in the current study and the ground state properties are expected to be close to the actual binding energy of these superheavy isotopes.

Separation energy of neutron shown in fig. 3.3(a) and (b), is positive for all the nuclei having $Z = 130 - 144$ within $290 \leq A \leq 380$, which indicates the neutron emission is hardly possible in these nuclei. The Q value associated with 1proton and 2proton decay is, $Q = -S_{1p}$ and $Q = -S_{2p}$. The one proton separation energy is illustrated in fig. 3.3 (c). The nuclei lying below the dashed line $S_{1p} = 0$ are the possible proton emitters and those that lie above are stable with respect to proton decay, similar to S_{2p} as in fig. 3.3 (d). The spacing between $Z = 138$ and $Z = 140$ separation energy curve is wider than the gap between any other successive Z separation energy curve indicates the existence of proton shell closure at $Z = 138$, which coincides with the previous works (Patra et al., 1999).

c) Decay half-life

For the isotopes of $Z = 130 - 144$, the Q values, half-lives of proton emission, alpha decay and cluster decay are calculated and tabulated in table 3.2 - 3.9 and are compared with theoretical

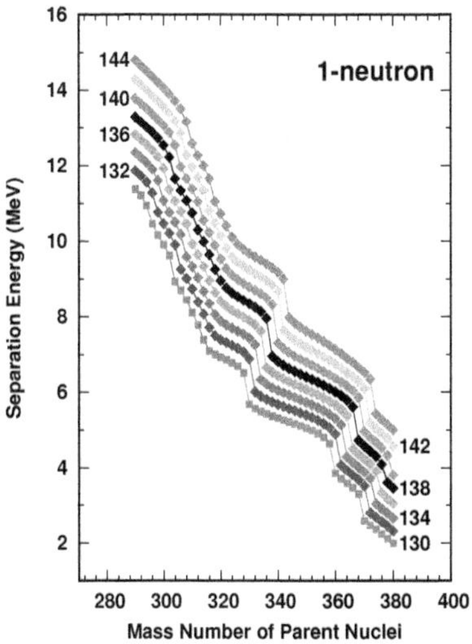

(a) one neutron separation energy

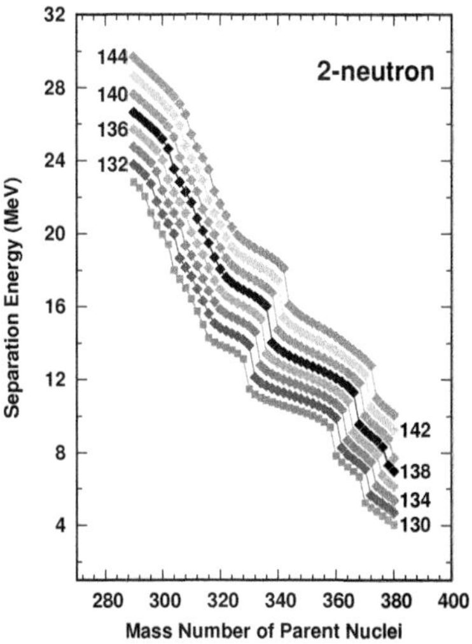

(b) two neutron separation energy

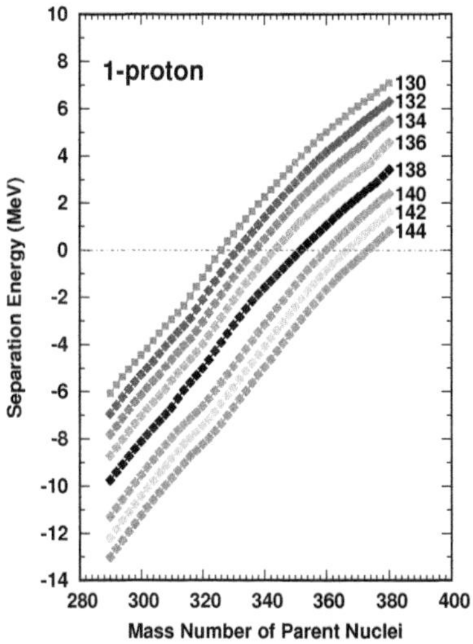

(c) one proton separation energy

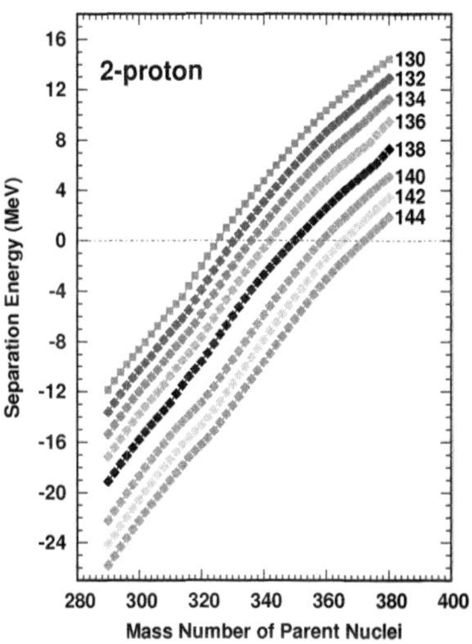

(d) two proton separation energy

Figure 3.3: The Binding energy of DD-PC1 parametrization for (a) 1-neutron , (b) 2-neutron , (c) 1-proton and (d) 2-proton plotted against the parent nuclei mass number for Z=130-144.

approaches obtained from Ni.et al., (Ni and Ren, 2012) for proton emission, Universal decay law(Qi et al., 2009) and Royer et al., (Royer, 2000) for alpha decay and Universal decay law (Qi et al., 2009) for cluster decay. The theoretical decay properties of heavy and superheavy isotopes have been widely investigated by different authors, wherein alpha emission, emission of light clusters such as ^{8}Be, ^{10}Be, ^{12}C, ^{14}C, ^{16}O, ^{20}O, and other heavy clusters have been studied (Santhosh and Sahadevan, 2010; Santhosh and Nithya, 2017; Manjunatha and Sowmya, 2018; Manjunatha et al., 2018). In this work, we limit to examine the decay properties of very light clusters of alpha multiples with $N = Z$ and in this context, the light clusters ^{8}Be and ^{12}C are considered.

The one-proton emission half-life is calculated using the theoretical approach described in chapter 2, is in good agreement with the empirical formula of Ni et al., (Ni and Ren, 2012), and is illustrated in table 3.2 - 3.9. The half-life calculation on the present approach is presented in fig. 3.4. From the plot 3.4 it is evident that the isotopes of $Z = 138$ are relatively more stable due to its large spacing gap with respect to neighbouring isotopes $Z = 140$, calculated using improved transfer matrix method which favours the prediction of proton shell closure at $Z = 138$ by separation energy calculation fig. 3.3(c) & (d). The shell closure at $Z = 132$ suggested by Zhang et al., (W. Zhang et al., 2005) is not seen with our DD-PC1 calculations.

In fig. 3.5 - 3.7 the Q-values plotted against the mass number of parent show dips at $N = 172, 184, 198, 228$ and 238 for alpha and cluster decays and suggest magicity at these places. Santhosh et al., (Santhosh and Nithya, 2017; Santhosh and Priyanka, 2017) studied the alpha decay and cluster emission at $Z = 110, 122, 126, 128$ and suggested $N = 184$ and $N = 202$ as the magic numbers using data from Moller et al., (Moller et al., 2016) and WS4 mass table (Wang et al., 2014). Our current study with $Z = 130 - 144$ isotopes brings out more shell closures at $N = 172, 184, 198, 228$ and 238. Such a consistent trend seen is the evidence for new neutron magic numbers. Hence $N = 172, 184, 198, 228$ and 238 are likely to be the neutron magic numbers which have been previously reported for different relativistic mean field parametrization (Gambhir et al., 2009; Bhuyan and Patra, 2012).

In fig. 3.8 - 3.10 the half-life obtained with the present calculation is plotted w.r.t. parent nuclei neutron number. The sudden peak in half-life curves is seen for to all Z at $N = 198, 228$

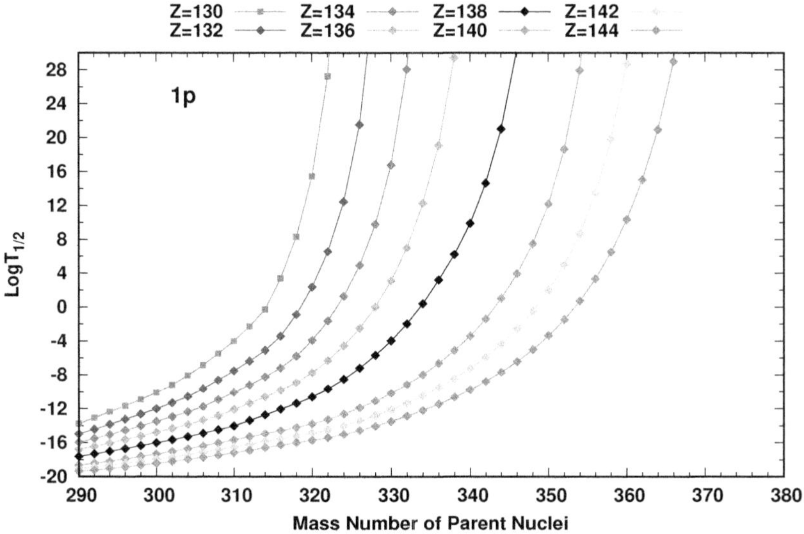

Figure 3.4: Half lives for 1proton emission against mass number of Parent nuclei for $Z = 130 - 144$. Comparatively large gap seen between $Z = 138$ and $Z = 140$ predicts the proton shell closure.

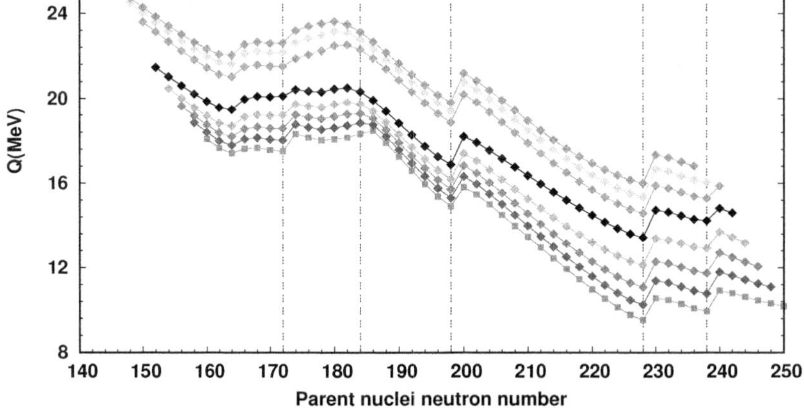

Figure 3.5: Comparison of Q-value with the parent nuclei neutron number for ${}_2^4He$ decay of $Z = 130 - 144$. The uniform dip seen coincides at the predicted neutron magic numbers.

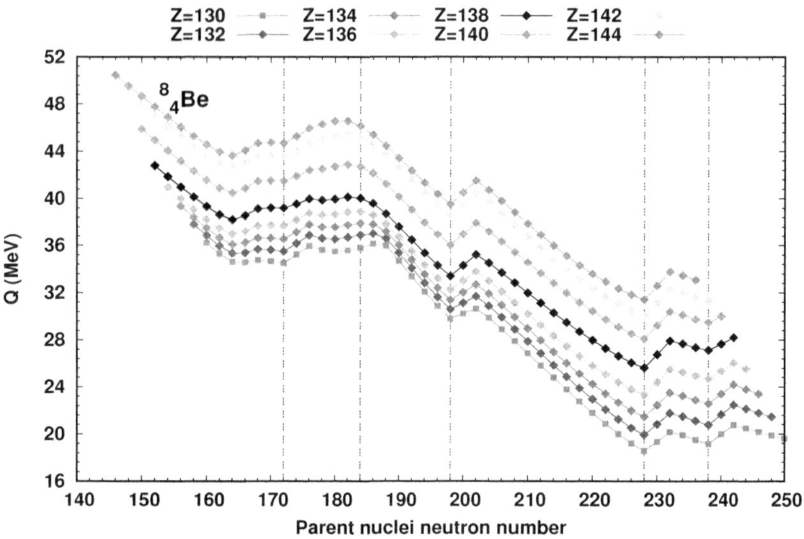

Figure 3.6: Same as in fig. 3.5 but for ${}^{8}_{4}Be$ decay.

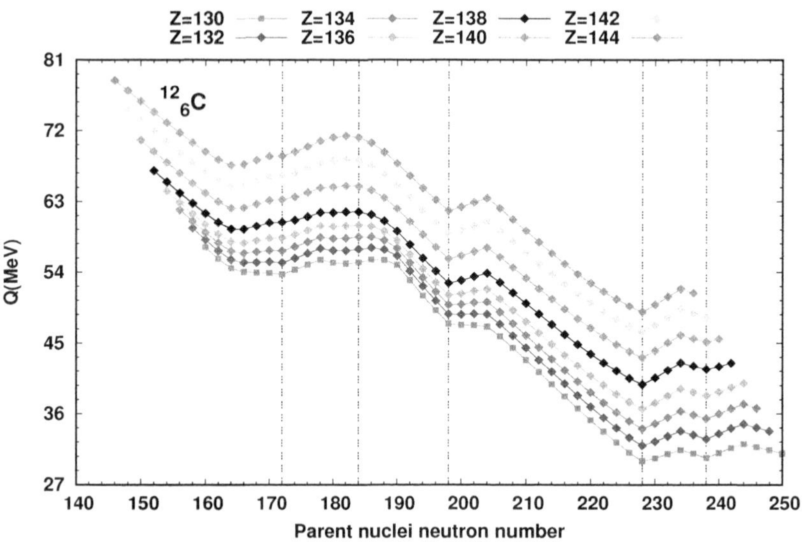

Figure 3.7: Same as in fig. 3.5 but for ${}^{12}_{6}C$ decay.

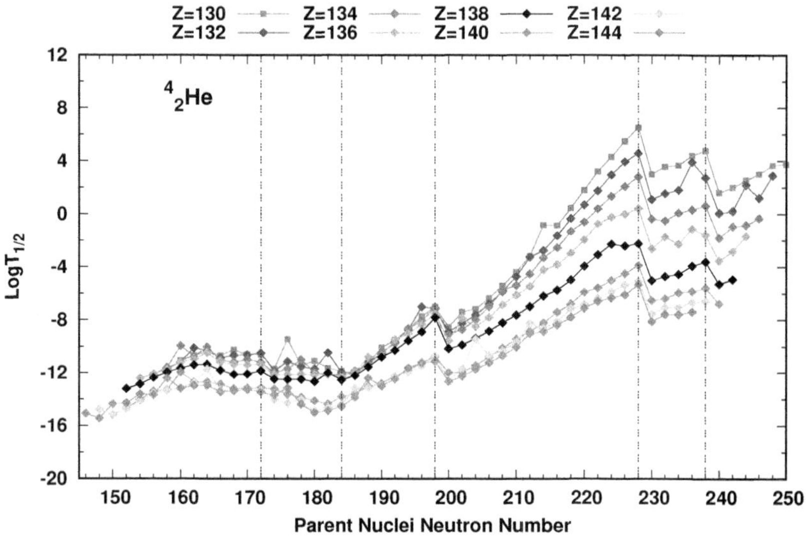

Figure 3.8: Half lives for ^4_2He emission against neutron number of Parent nuclei for $Z = 130-144$. The vertical dotted lines indicates higher stability against alpha emission.

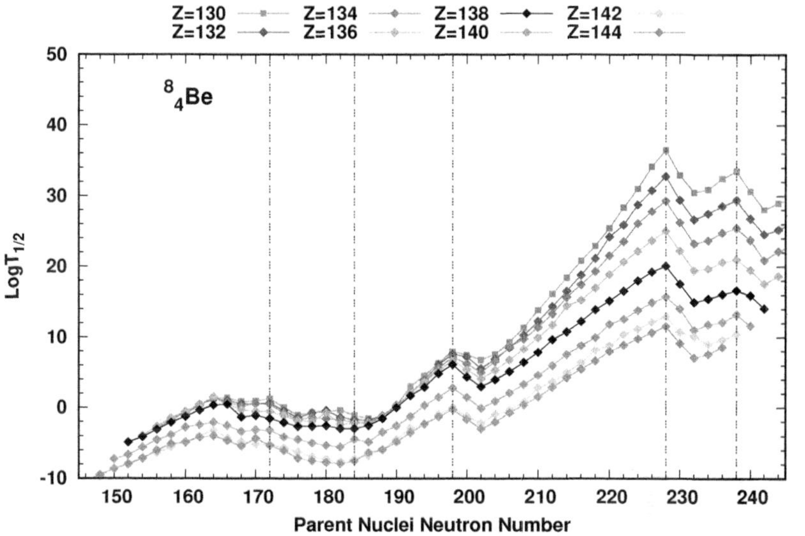

Figure 3.9: Same as in fig. 3.8 but for ^8_4Be decay.

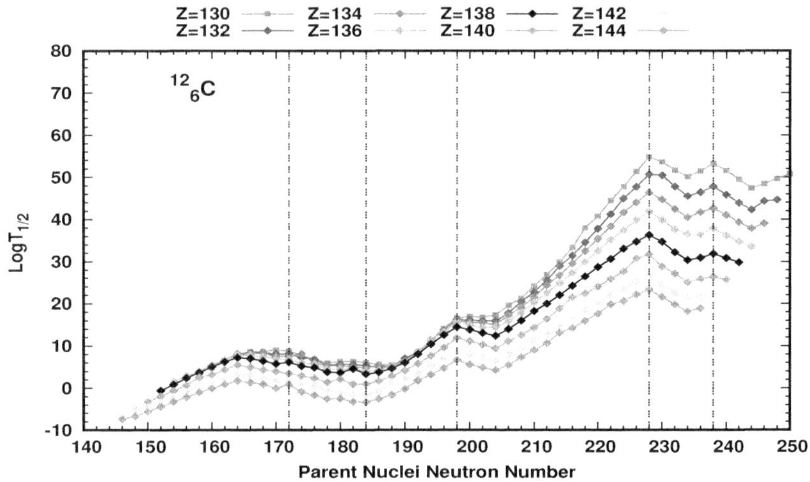

Figure 3.10: Same as in fig. 3.8 but for $^{12}_{6}C$ decay.

and 238; the alpha emission in some parent nuclei happens to be relatively longer and more stable, which suggests neutron shell closure and magicity. At $N = 172$ and 184, the peaks tend to become smooth for the charge number being dealt is higher. A similar trend of peaks in half-life curves is seen at $N = 172, 184, 198, 228$ and 238, for cluster decay (Be , C) (fig. 3.9 and 3.10) which provides conclusive evidence that emission of any fragment α, Be or C happens to be relatively longer in some parent nuclei attaining shell closure. From fig. 3.4 a wider gap between $Z = 138$ and $Z = 140$ as compared to other successive Z values observed in proton emission is also present in α decay and Be decay half-life curves, whereas for C decay half-life curves the gap between $Z = 138$ and $Z = 140$ is less pronounced. Such peaks and the wide gap in half-life curves as in fig. 3.4, 3.8 - 3.10 are the signatures of magicity.

The current experimental constraints limit the observation of decay with half-life between $10^{-6} \leq T_{1/2} \leq 10^{30}$ s (Santhosh and Priyanka, 2017; Santhosh and Nithya, 2017). The competition between various decay modes such as one-proton emission, alpha and light-cluster decay and SF are illustrated in fig. 3.11 and 3.12 where the SF half-lives are evaluated with the formula of Xu et al., (Xu and Ren, 2005). The dashed lines correspond to $T_{1/2} = 10^{-6}$ and $T_{1/2} = 10^{30}$s, which encloses the decay modes associated with nuclei which are in current experimental observation limitation.

Table 3.2: The Q-value (MeV) and the half lives ($logT_{1/2}; T_{1/2}$ in s) for proton emission, alpha decay, cluster emission of ${}^{8}_{4}Be$ and ${}^{12}_{6}C$ and SF for $Z = 130$ even-even isotopes.

Z=130	Proton decay			Alpha decay				${}^{8}_{4}Be$ Decay			${}^{12}_{6}C$ Decay			Fission
A	Q_p	$logT_p$	Ni.	Q_α	$logT_\alpha$	UDL	Royer	Q_{Be}	$logT_{Be}$	UDL	Q_C	$logT_C$	UDL	Xu
290	6.06	-13.76	-13.32	18.08	-11.07	-11.32	-10.92	36.23	-0.88	0.11	57.16	4.90	5.96	17.79
292	5.71	-13.09	-12.67	17.66	-10.69	-10.75	-10.38	35.32	0.33	1.34	55.74	6.66	7.68	26.01
294	5.36	-12.37	-11.98	17.41	-10.38	-10.41	-10.07	34.63	1.35	2.28	54.56	7.99	9.15	33.51
296	5.05	-11.66	-11.28	17.62	-10.66	-10.76	-10.41	34.57	1.40	2.31	54.05	8.67	9.75	40.30
298	4.75	-10.90	-10.55	17.67	-10.24	-10.86	-10.51	34.75	0.89	1.98	53.94	8.69	9.82	46.37
300	4.45	-10.09	-9.75	17.57	-10.59	-10.74	-10.41	34.67	0.94	2.04	53.92	9.09	9.75	51.73
302	4.15	-9.18	-8.89	17.51	-10.61	-10.69	-10.37	34.49	1.28	2.23	53.73	8.77	9.91	56.37
304	3.83	-8.06	-7.82	18.32	-11.69	-11.90	-11.51	35.23	0.06	1.09	54.34	7.78	8.99	60.29
306	3.52	-6.84	-6.65	18.16	-9.44	-11.70	-11.33	35.94	-1.05	0.01	54.95	6.88	8.07	63.50
308	3.21	-5.46	-5.36	18.03	-11.41	-11.55	-11.20	35.62	-0.68	0.40	55.57	5.97	7.16	65.99
310	2.92	-3.99	-3.95	18.07	-11.07	-11.64	-11.29	35.50	-0.61	0.51	55.23	6.34	7.52	67.77
312	2.65	-2.28	-2.37	18.16	-11.63	-11.80	-11.44	35.58	-0.35	0.33	55.11	6.37	7.59	68.83
314	2.36	-0.28	-0.46	18.33	-11.85	-12.07	-11.70	35.81	-1.10	-0.05	55.26	6.03	7.30	69.18
316	1.95	3.38	2.95	18.48	-12.05	-12.32	-11.94	36.12	-1.53	-0.56	55.58	5.60	6.79	68.81
318	1.55	8.33	7.54	17.90	-11.24	-11.52	-11.20	35.98	-1.53	-0.41	55.53	5.48	6.77	67.73
320	1.16	15.44	14.18	17.24	-10.08	-10.58	-10.32	34.69	0.25	1.38	54.93	6.21	7.48	65.94
322	0.77	27.25	25.18	16.59	-9.52	-9.58	-9.39	33.37	3.04	3.34	52.98	8.64	10.07	63.43
324	0.39	53.37	49.51	15.96	-8.72	-8.56	-8.44	32.08	4.52	5.38	51.03	11.31	12.82	60.21
326	0.01	463.29	440.31	15.37	-7.73	-7.55	-7.50	30.85	6.30	7.43	49.15	14.11	15.63	56.27
328	-0.37	-	-	14.91	-7.02	-6.73	-6.73	29.80	7.98	9.29	47.46	16.62	18.30	51.63
330	-0.68	-	-	15.81	-8.54	-8.38	-8.29	30.23	7.48	8.43	47.30	16.97	18.48	46.27
332	-1.00	-	-	15.47	-7.39	-7.81	-7.76	30.65	6.78	7.62	47.27	16.74	18.43	40.19
334	-1.32	-	-	15.00	-7.18	-6.98	-6.99	29.86	7.58	8.99	47.10	17.27	18.64	33.41
336	-1.64	-	-	14.50	-6.36	-6.05	-6.12	28.90	9.32	10.76	45.85	19.60	20.72	25.91
338	-1.95	-	-	13.99	-5.41	-5.03	-5.17	27.89	11.36	12.74	44.38	21.28	23.27	17.70
340	-2.25	-	-	13.47	-4.39	-3.94	-4.15	26.86	13.86	14.87	42.85	24.17	26.09	8.78
342	-2.56	-	-	12.96	-3.19	-2.79	-3.08	25.82	16.13	17.15	41.30	26.89	29.11	-0.85
344	-2.86	-	-	12.45	-0.83	-1.58	-1.95	24.79	18.44	19.56	39.75	29.96	32.33	-11.19
346	-3.15	-	-	11.94	-0.84	-0.32	-0.77	23.77	20.79	22.10	38.20	33.38	35.72	-22.25
348	-3.44	-	-	11.46	0.45	0.99	0.45	22.77	22.92	24.76	36.67	38.00	39.28	-34.01
350	-3.73	-	-	10.99	1.80	2.33	1.71	21.79	25.41	27.51	35.18	40.81	42.99	-46.49
352	-4.00	-	-	10.54	3.20	3.68	2.97	20.86	28.40	30.33	33.73	44.46	46.82	-59.67
354	-4.27	-	-	10.13	4.31	5.00	4.21	19.99	31.00	33.16	32.36	47.72	50.70	-73.57
356	-4.54	-	-	9.78	5.50	6.23	5.35	19.19	34.15	35.90	31.08	51.32	54.54	-88.18
358	-4.80	-	-	9.53	6.52	7.09	6.15	18.56	36.50	38.18	29.99	54.75	57.99	-103.50
360	-5.02	-	-	10.57	2.98	3.48	2.76	19.33	32.97	35.29	30.34	53.61	56.74	-119.52
362	-5.24	-	-	10.48	3.58	3.74	3.00	20.16	30.53	32.36	30.89	51.61	54.90	-136.26
364	-5.46	-	-	10.29	3.66	4.35	3.56	19.89	30.94	33.22	31.40	50.09	53.23	-153.71
366	-5.69	-	-	10.09	4.41	5.00	4.17	19.49	32.49	34.55	30.94	51.45	54.59	-171.87
368	-5.91	-	-	9.94	4.78	5.49	4.62	19.14	33.54	35.77	30.39	53.19	56.29	-190.73
370	-6.10	-	-	10.92	1.64	2.25	1.57	19.97	30.70	32.78	31.01	51.57	54.21	-210.31
372	-6.30	-	-	10.80	2.02	2.60	1.90	20.75	28.07	30.17	31.64	49.46	52.17	-230.59
374	-6.50	-	-	10.62	2.57	3.12	2.38	20.49	29.01	30.96	32.16	47.37	50.52	-251.59
376	-6.69	-	-	10.46	3.01	3.61	2.83	20.17	29.78	31.96	31.80	48.41	51.54	-273.29
378	-6.89	-	-	10.32	3.67	4.04	3.23	19.88	30.94	32.89	31.38	49.61	52.77	-295.70
380	-7.09	-	-	10.18	3.78	4.48	3.63	19.61	31.44	33.77	30.97	50.75	53.98	-318.82

Table 3.3: Same as in table 3.2 but for $Z = 132$ even-even isotopes.

Z=132	Proton decay			Alpha decay				$^{8}_{4}Be$ Decay			$^{12}_{6}C$ Decay			Fission
A	Q_p	$logT_p$	Ni.	Q_α	$logT_\alpha$	UDL	Royer	Q_{Be}	$logT_{Be}$	UDL	Q_C	$logT_C$	UDL	Xu
290	6.94	-14.97	-14.53	18.85	-11.69	-11.89	-11.46	37.81	-1.79	-1.04	59.61	3.47	4.33	14.70
292	6.58	-14.42	-14.00	18.40	-11.18	-11.30	-10.90	36.85	-0.58	0.19	58.13	5.24	6.05	24.39
294	6.24	-13.85	-13.43	18.00	-10.12	-10.76	-10.41	35.98	0.47	1.35	56.75	6.74	7.71	33.35
296	5.90	-13.24	-12.82	17.78	-10.31	-10.47	-10.14	35.34	1.17	2.21	55.62	8.12	9.10	41.60
298	5.59	-12.63	-12.23	18.07	-10.85	-10.93	-10.58	35.38	1.06	2.08	55.22	8.37	9.54	49.13
300	5.29	-12.00	-11.60	18.14	-10.66	-11.06	-10.71	35.66	0.55	1.61	55.23	8.42	9.44	55.95
302	4.99	-11.29	-10.92	18.06	-10.62	-10.97	-10.64	35.62	0.55	1.60	55.32	8.08	9.22	62.05
304	4.71	-10.56	-10.21	18.03	-10.50	-10.96	-10.63	35.49	0.63	1.73	55.21	8.13	9.28	67.43
306	4.38	-9.61	-9.30	18.77	-11.84	-12.04	-11.65	36.18	-0.40	0.68	55.78	7.20	8.42	72.09
308	4.06	-8.61	-8.34	18.63	-11.12	-11.89	-11.51	36.85	-1.39	-0.32	56.37	6.76	7.55	76.04
310	3.77	-7.52	-7.30	18.53	-11.53	-11.78	-11.42	36.58	-0.71	-0.02	56.98	5.45	6.67	79.28
312	3.48	-6.37	-6.20	18.60	-11.69	-11.92	-11.55	36.53	-0.40	0.00	56.74	5.68	6.89	81.80
314	3.20	-5.08	-4.97	18.71	-10.47	-12.10	-11.73	36.68	-1.38	-0.27	56.73	5.55	6.81	83.60
316	2.89	-3.42	-3.42	18.83	-11.96	-12.30	-11.93	36.89	-1.71	-0.63	56.93	5.33	6.46	84.69
318	2.50	-0.88	-1.04	18.76	-12.00	-12.23	-11.87	36.99	-1.78	-0.82	57.08	4.99	6.18	85.06
320	2.10	2.37	1.99	18.21	-11.14	-11.49	-11.18	36.59	-1.13	-0.33	56.85	5.10	6.39	84.72
322	1.71	6.55	5.95	17.58	-10.59	-10.59	-10.34	35.37	0.32	1.35	56.07	7.03	7.32	83.67
324	1.33	12.45	11.43	16.94	-9.66	-9.64	-9.45	34.09	2.02	3.22	54.15	8.52	9.82	81.90
326	0.95	21.50	19.82	16.33	-8.74	-8.66	-8.54	32.82	3.92	5.18	52.22	10.99	12.50	79.42
328	0.58	37.92	35.17	15.76	-7.00	-7.70	-7.64	31.62	6.13	7.15	50.35	13.82	15.23	76.22
330	0.21	87.94	82.14	15.30	-7.13	-6.90	-6.90	30.58	7.55	8.95	48.67	16.19	17.82	72.32
332	-0.11	-	-	16.31	-8.89	-8.72	-8.62	31.12	7.22	7.90	48.62	16.12	17.82	67.70
334	-0.44	-	-	15.96	-8.25	-8.15	-8.09	31.67	5.52	6.89	48.71	15.86	17.59	62.36
336	-0.75	-	-	15.50	-7.62	-7.35	-7.35	30.86	6.88	8.25	48.66	15.87	17.59	56.32
338	-1.07	-	-	15.00	-6.78	-6.46	-6.51	29.91	8.50	9.96	47.39	17.78	19.63	49.56
340	-1.38	-	-	14.50	-5.85	-5.49	-5.61	28.90	10.30	11.85	45.92	20.41	22.10	42.09
342	-1.69	-	-	13.99	-4.74	-4.47	-4.66	27.88	12.26	13.87	44.40	22.76	24.79	33.90
344	-1.99	-	-	13.49	-3.28	-3.39	-3.65	26.86	14.34	16.02	42.86	25.52	27.67	25.01
346	-2.28	-	-	12.99	-2.75	-2.27	-2.61	25.84	16.52	18.28	41.33	29.00	30.71	15.40
348	-2.57	-	-	12.51	-1.64	-1.12	-1.53	24.85	18.78	20.63	39.81	31.45	33.89	5.09
350	-2.86	-	-	12.04	-0.35	0.06	-0.43	23.88	21.14	23.06	38.32	34.55	37.19	-5.94
352	-3.13	-	-	11.59	0.68	1.25	0.69	22.95	24.19	25.54	36.88	37.82	40.59	-17.68
354	-3.40	-	-	11.18	1.75	2.43	1.78	22.07	25.90	28.04	35.49	41.25	44.05	-30.13
356	-3.66	-	-	10.80	2.96	3.56	2.84	21.24	28.74	30.51	34.18	44.95	47.51	-43.29
358	-3.92	-	-	10.47	3.94	4.58	3.79	20.51	30.78	32.84	32.98	47.68	50.87	-57.17
360	-4.17	-	-	10.26	4.57	5.27	4.43	19.93	32.79	34.73	31.97	50.68	53.82	-71.75
362	-4.39	-	-	11.38	1.09	1.71	1.09	20.82	29.45	31.71	32.46	50.43	52.24	-87.04
364	-4.61	-	-	11.29	1.55	1.96	1.31	21.76	26.63	28.69	33.14	47.75	50.17	-103.04
366	-4.82	-	-	11.10	1.81	2.48	1.80	21.48	27.48	29.49	33.79	45.50	48.24	-119.75
368	-5.04	-	-	10.91	3.92	3.04	2.32	21.10	28.58	30.65	33.32	46.40	49.48	-137.18
370	-5.26	-	-	10.77	2.75	3.44	2.69	20.76	29.44	31.69	32.79	47.80	50.97	-155.31
372	-5.47	-	-	11.79	0.08	0.42	-0.16	21.63	26.79	28.88	33.45	45.83	48.96	-174.15
374	-5.68	-	-	11.63	0.23	0.85	0.24	22.45	24.55	26.39	34.12	43.91	47.00	-193.70
376	-5.89	-	-	11.43	2.20	1.38	0.73	22.12	25.20	27.28	34.70	42.25	45.33	-213.96
378	-6.10	-	-	11.25	1.21	1.87	1.19	21.77	26.19	28.28	34.26	44.31	46.46	-234.93
380	-6.30	-	-	11.10	2.91	2.30	1.59	21.46	27.12	29.20	33.79	44.61	47.70	-256.60

Table 3.4: Same as in table 3.2 but for $Z = 134$ even-even isotopes.

Z=134	Proton decay			Alpha decay				$^{8}_{4}Be$ Decay			$^{12}_{6}C$ Decay			Fission
A	Q_p	$logT_p$	Ni.	Q_α	$logT_\alpha$	UDL	Royer	Q_{Be}	$logT_{Be}$	UDL	Q_C	$logT_C$	UDL	Xu
290	7.81	-15.95	-15.54	19.64	-12.13	-12.45	-11.98	39.38	-2.84	-2.11	61.99	2.33	2.89	10.63
292	7.46	-15.51	-15.09	19.18	-11.60	-11.87	-11.44	38.42	-1.83	-0.93	60.51	3.98	4.53	21.78
294	7.12	-15.04	-14.61	18.75	-9.94	-11.31	-10.92	37.50	-0.62	0.23	59.08	5.43	6.18	32.21
296	6.78	-14.54	-14.12	18.38	-10.77	-10.82	-10.47	36.67	0.35	1.31	57.75	6.74	7.75	41.92
298	6.45	-14.02	-13.59	18.20	-10.03	-10.59	-10.27	36.10	1.54	2.07	56.69	8.03	9.03	50.91
300	6.15	-13.49	-13.07	18.56	-11.01	-11.15	-10.79	36.24	0.77	1.79	56.42	8.16	9.30	59.19
302	5.85	-12.94	-12.53	18.65	-11.16	-11.31	-10.95	36.62	0.24	1.19	56.55	8.30	9.03	66.75
304	5.56	-12.35	-11.95	18.58	-10.99	-11.25	-10.90	36.62	0.55	1.12	56.76	7.50	8.66	73.59
306	5.28	-11.73	-11.35	18.58	-11.18	-11.28	-10.94	36.54	0.48	1.18	56.72	7.50	8.62	79.71
308	4.94	-10.91	-10.55	19.22	-11.99	-12.20	-11.81	37.15	-0.75	0.26	57.23	8.14	7.86	85.12
310	4.62	-10.07	-9.73	19.11	-11.69	-12.08	-11.70	37.78	-1.71	-0.66	57.80	5.84	7.03	89.81
312	4.32	-9.19	-8.87	19.04	-10.98	-12.01	-11.65	37.56	-1.39	-0.44	58.40	5.41	6.18	93.79
314	4.04	-8.22	-7.97	19.14	-11.95	-12.19	-11.82	37.57	-1.58	-0.51	58.25	5.26	6.28	97.05
316	3.75	-7.17	-6.96	19.26	-12.13	-12.38	-12.00	37.76	-1.70	-0.83	58.31	5.03	6.12	99.59
318	3.42	-5.79	-5.66	19.29	-12.11	-12.45	-12.08	37.90	-2.09	-1.08	58.49	4.76	5.79	101.42
320	3.04	-3.91	-3.88	19.06	-11.83	-12.17	-11.82	37.79	-2.07	-0.99	58.47	5.39	5.74	102.54
322	2.66	-1.61	-1.69	18.52	-11.37	-11.45	-11.15	37.18	-1.13	-0.22	58.02	4.99	6.21	102.94
324	2.27	1.29	0.99	17.91	-10.43	-10.60	-10.36	36.02	0.29	1.36	57.02	6.12	7.41	102.62
326	1.90	4.93	4.40	17.30	-9.76	-9.70	-9.52	34.78	2.04	3.14	55.19	8.50	9.77	101.59
328	1.52	9.79	8.94	16.71	-8.63	-8.77	-8.66	33.55	3.75	5.01	53.31	10.93	12.33	99.85
330	1.16	16.75	15.42	16.16	-8.04	-7.86	-7.81	32.38	5.58	6.88	51.49	13.82	14.95	97.40
332	0.79	28.07	25.95	15.71	-7.06	-7.11	-7.11	31.37	7.17	8.59	49.85	16.03	17.43	94.23
334	0.47	47.57	44.15	16.82	-9.09	-9.06	-8.94	32.03	6.23	7.38	49.92	15.52	17.23	90.35
336	0.15	113.56	105.88	16.47	-8.72	-8.51	-8.43	32.68	4.99	6.20	50.12	15.43	16.83	85.75
338	-0.17	-	-	16.02	-7.97	-7.75	-7.73	31.88	6.55	7.51	50.21	14.98	16.61	80.44
340	-0.48	-	-	15.54	-6.99	-6.91	-6.95	30.94	8.65	9.13	48.92	17.23	18.60	74.42
342	-0.79	-	-	15.05	-5.76	-6.01	-6.11	29.95	9.77	10.92	47.47	19.08	20.96	67.69
344	-1.09	-	-	14.56	-5.40	-5.07	-5.23	28.96	11.42	12.81	45.98	21.56	23.51	60.25
346	-1.38	-	-	14.07	-4.54	-4.08	-4.31	27.96	13.34	14.81	44.47	24.33	26.22	52.09
348	-1.67	-	-	13.60	-3.31	-3.07	-3.37	26.98	15.65	16.88	42.97	26.81	29.06	43.22
350	-1.95	-	-	13.14	-2.55	-2.04	-2.41	26.03	17.43	19.02	41.50	29.60	32.00	33.64
352	-2.22	-	-	12.71	-1.32	-1.01	-1.44	25.11	19.31	21.19	40.07	32.49	35.02	23.35
354	-2.49	-	-	12.29	-0.62	0.01	-0.49	24.23	21.55	23.38	38.70	35.43	38.08	12.35
356	-2.75	-	-	11.91	0.39	1.01	0.44	23.41	23.54	25.55	37.38	38.37	41.16	0.63
358	-3.00	-	-	11.57	1.34	1.95	1.31	22.65	26.05	27.65	36.16	41.64	44.18	-11.79
360	-3.24	-	-	11.28	2.10	2.78	2.08	21.98	27.78	29.60	35.04	44.02	47.07	-24.93
362	-3.48	-	-	11.09	2.80	3.32	2.59	21.46	29.34	31.14	34.12	46.37	49.56	-38.77
364	-3.70	-	-	12.29	-0.38	-0.11	-0.64	22.45	26.22	28.08	34.74	44.70	47.73	-53.33
366	-3.91	-	-	12.20	-0.50	0.09	-0.45	23.48	23.25	25.07	35.53	42.44	45.50	-68.60
368	-4.13	-	-	12.03	0.09	0.53	-0.05	23.22	23.69	25.76	36.30	40.47	43.41	-84.57
370	-4.34	-	-	11.85	0.34	0.98	0.37	22.86	24.80	26.73	35.85	41.71	44.51	-101.26
372	-4.53	-	-	11.74	0.63	1.25	0.62	22.56	25.43	27.55	35.36	42.64	45.74	-118.66
374	-4.79	-	-	12.70	-1.82	-1.32	-1.80	23.38	23.78	25.15	35.98	41.06	44.02	-136.76
376	-5.03	-	-	12.49	-0.93	-0.80	-1.32	24.18	20.88	22.93	36.63	39.35	42.25	-155.58
378	-5.26	-	-	12.26	-0.81	-0.24	-0.80	23.79	22.17	23.92	37.23	37.84	40.69	-175.10
380	-5.48	-	-	12.06	-0.32	0.26	-0.34	23.39	22.98	24.95	36.71	39.05	41.92	-195.34

Table 3.5: Same as in table 3.2 but for $Z = 136$ even-even isotopes.

Z=136	Proton decay			Alpha decay				$^{8}_{4}Be$ Decay			$^{12}_{6}C$ Decay			Fission
A	Q_p	$logT_p$	Ni.	Q_α	$logT_\alpha$	UDL	Royer	Q_{Be}	$logT_{Be}$	UDL	Q_C	$logT_C$	UDL	Xu
290	8.71	-16.80	-16.42	20.47	-12.41	-13.01	-12.51	40.98	-3.96	-3.16	64.37	1.51	1.53	5.67
292	8.36	-16.43	-16.04	20.01	-12.08	-12.47	-12.01	40.03	-2.67	-2.05	62.90	2.93	3.08	18.29
294	8.02	-16.05	-15.64	19.57	-11.75	-11.93	-11.51	39.11	-1.47	-0.95	61.47	3.90	4.65	30.18
296	7.69	-15.64	-15.22	19.17	-11.17	-11.43	-11.04	38.25	-0.54	0.13	60.11	5.28	6.20	41.35
298	7.36	-15.22	-14.78	18.84	-10.94	-11.00	-10.65	37.48	0.41	1.11	58.85	6.71	7.66	51.80
300	7.04	-14.77	-14.33	18.70	-10.45	-10.84	-10.50	36.97	1.41	1.76	57.89	7.76	8.80	61.54
302	6.75	-14.32	-13.88	19.12	-11.16	-11.47	-11.10	37.22	0.54	1.34	57.74	7.81	8.91	70.56
304	6.46	-13.85	-13.41	19.23	-11.35	-11.65	-11.28	37.69	-0.29	0.62	57.99	7.44	8.48	78.86
306	6.17	-13.34	-12.92	19.19	-11.37	-11.62	-11.26	37.73	-0.52	0.50	58.33	7.01	7.96	86.44
308	5.89	-12.83	-12.40	19.22	-11.34	-11.70	-11.34	37.70	-0.52	0.49	58.36	6.69	7.82	93.31
310	5.54	-12.08	-11.68	19.72	-12.15	-12.40	-12.01	38.20	-1.12	-0.26	58.78	6.06	7.20	99.46
312	5.21	-11.35	-10.96	19.64	-12.11	-12.32	-11.94	38.76	-1.85	-1.07	59.30	5.62	6.44	104.90
314	4.91	-10.59	-10.23	19.59	-12.08	-12.29	-11.92	38.60	-1.94	-0.92	59.88	4.50	5.63	109.61
316	4.62	-9.80	-9.47	19.72	-12.14	-12.50	-12.12	38.65	-0.75	-1.05	59.80	4.48	5.64	113.62
318	4.32	-8.91	-8.61	19.82	-12.12	-12.66	-12.27	38.86	-2.34	-1.39	59.90	4.62	5.42	116.90
320	3.98	-7.76	-7.50	19.74	-12.27	-12.60	-12.22	38.90	-2.22	-1.50	60.02	4.01	5.19	119.47
322	3.61	-6.28	-6.11	19.40	-11.93	-12.17	-11.83	38.59	-2.24	-1.16	59.82	4.23	5.34	121.33
324	3.23	-4.53	-4.46	18.87	-10.78	-11.48	-11.19	37.83	-1.30	-0.19	59.18	5.28	6.05	122.47
326	2.85	-2.50	-2.52	18.29	-10.71	-10.68	-10.44	36.71	0.26	1.31	57.99	6.29	7.48	122.90
328	2.48	0.00	-0.18	17.71	-9.47	-9.83	-9.65	35.52	2.05	2.99	56.24	8.34	9.70	122.61
330	2.12	3.10	2.70	17.14	-9.05	-8.96	-8.84	34.34	3.70	4.75	54.43	11.54	12.13	121.61
332	1.76	7.00	6.36	16.61	-8.31	-8.12	-8.06	33.22	5.19	6.51	52.68	13.20	14.61	119.89
334	1.41	12.27	11.29	16.19	-7.52	-7.42	-7.41	32.24	6.70	8.11	51.09	15.61	16.95	117.46
336	1.09	19.06	17.61	17.40	-9.57	-9.47	-9.34	33.01	5.38	6.74	51.28	14.89	16.56	114.32
338	0.78	29.40	27.34	17.06	-7.89	-8.96	-8.87	33.78	4.08	5.40	51.60	14.32	15.98	110.46
340	0.47	48.64	45.21	16.63	-8.49	-8.28	-8.23	33.00	5.30	6.63	51.83	14.35	15.55	105.90
342	0.17	105.79	98.78	16.17	-7.80	-7.52	-7.53	32.10	6.75	8.13	50.57	15.82	17.43	100.61
344	-0.12	-	-	15.71	-6.85	-6.72	-6.78	31.15	8.31	9.76	49.17	17.76	19.63	94.62
346	-0.41	-	-	15.26	-6.13	-5.89	-6.01	30.21	9.94	11.48	47.73	20.27	21.99	87.91
348	-0.69	-	-	14.81	-5.46	-5.04	-5.22	29.27	11.74	13.26	46.28	22.53	24.47	80.49
350	-0.96	-	-	14.38	-4.23	-4.18	-4.42	28.35	14.45	15.09	44.86	24.88	27.04	72.36
352	-1.22	-	-	13.97	-3.82	-3.32	-3.62	27.47	15.24	16.95	43.47	27.37	29.67	63.52
354	-1.48	-	-	13.58	-2.99	-2.47	-2.82	26.62	16.97	18.81	42.12	29.94	32.35	53.97
356	-1.73	-	-	13.21	-1.97	-1.64	-2.05	25.81	18.88	20.66	40.84	32.56	35.02	43.70
358	-1.97	-	-	12.87	-0.74	-0.84	-1.31	25.06	20.67	22.47	39.62	35.23	37.68	32.72
360	-2.21	-	-	12.56	-0.23	-0.09	-0.61	24.38	22.14	24.20	38.49	37.42	40.25	21.03
362	-2.45	-	-	12.30	0.01	0.57	0.01	23.77	23.65	25.78	37.47	39.86	42.68	8.64
364	-2.69	-	-	12.13	0.40	1.00	0.40	23.31	25.01	27.01	36.63	41.89	44.73	-4.47
366	-2.90	-	-	13.39	-2.64	-2.22	-2.62	24.38	22.20	24.02	37.37	39.89	42.76	-18.30
368	-3.11	-	-	13.31	-1.75	-2.05	-2.47	25.50	19.49	21.11	38.27	37.60	40.46	-32.83
370	-3.32	-	-	13.15	-2.28	-1.70	-2.15	25.25	19.70	21.68	39.15	36.52	38.29	-48.07
372	-3.53	-	-	12.99	-1.12	-1.35	-1.83	24.92	20.67	22.46	38.73	36.42	39.21	-64.02
374	-3.72	-	-	12.93	-1.59	-1.21	-1.70	24.68	21.04	23.02	38.30	37.97	40.16	-80.68
376	-4.02	-	-	13.70	-3.53	-3.07	-3.45	25.34	19.55	21.27	38.77	36.16	38.95	-98.05
378	-4.30	-	-	13.44	-2.81	-2.50	-2.92	26.04	17.56	19.51	39.33	34.76	37.55	-116.13
380	-4.56	-	-	13.17	-1.67	-1.89	-2.36	25.56	18.69	20.64	39.87	33.54	36.25	-134.92

Table 3.6: Same as in table 3.2 but for $Z = 138$ even-even isotopes.

Z=138	Proton decay		Alpha decay					${}^{8}_{4}Be$ Decay			${}^{12}_{6}C$ Decay			Fission
A	Q_p	$logT_p$	Ni.	Q_α	$logT_\alpha$	UDL	Royer	Q_{Be}	$logT_{Be}$	UDL	Q_C	$logT_C$	UDL	Xu
290	9.74	-17.62	-17.30	21.47	-13.22	-13.75	-13.21	42.79	-4.90	-4.38	66.93	-0.57	0.05	-0.08
292	9.40	-17.32	-16.98	21.03	-12.87	-13.26	-12.76	41.87	-4.13	-3.37	65.50	0.83	1.48	14.00
294	9.07	-17.01	-16.65	20.61	-12.36	-12.78	-12.31	40.97	-3.09	-2.35	64.11	2.40	2.93	27.36
296	8.75	-16.70	-16.31	20.21	-11.94	-12.32	-11.88	40.12	-2.08	-1.36	62.76	3.68	4.38	39.99
298	8.43	-16.36	-15.96	19.86	-11.64	-11.89	-11.49	39.33	-1.26	-0.40	61.49	5.06	5.79	51.91
300	8.12	-16.01	-15.60	19.57	-11.40	-11.54	-11.17	38.64	-0.35	0.45	60.33	6.21	7.11	63.11
302	7.82	-15.65	-15.22	19.48	-11.36	-11.45	-11.09	38.21	0.37	0.96	59.48	7.20	8.08	73.59
304	7.53	-15.28	-14.85	19.96	-11.80	-12.13	-11.73	38.56	0.49	0.42	59.45	7.00	8.02	83.35
306	7.25	-14.89	-14.45	20.10	-12.11	-12.34	-11.94	39.12	-1.32	-0.39	59.83	6.41	7.45	92.40
308	6.96	-14.48	-14.03	20.07	-12.08	-12.34	-11.94	39.20	-1.12	-0.56	60.28	5.75	6.80	100.72
310	6.68	-14.04	-13.58	20.11	-11.84	-12.42	-12.03	39.20	-1.54	-0.62	60.36	6.16	6.60	108.33
312	6.30	-13.38	-12.94	20.42	-12.47	-12.85	-12.44	39.54	-2.11	-1.13	60.63	5.21	6.17	115.23
314	5.94	-12.71	-12.28	20.34	-12.50	-12.79	-12.38	39.96	-2.65	-1.74	61.06	4.87	5.56	121.40
316	5.61	-12.05	-11.63	20.30	-12.48	-12.77	-12.38	39.84	-2.64	-1.65	61.58	3.79	4.83	126.86
318	5.31	-11.37	-10.98	20.45	-12.66	-12.99	-12.59	39.94	-2.56	-1.83	61.57	3.66	4.76	131.61
320	5.00	-10.62	-10.23	20.50	-12.01	-13.09	-12.69	40.12	-2.97	-2.13	61.67	4.56	4.54	135.63
322	4.65	-9.64	-9.30	20.32	-12.54	-12.90	-12.51	40.04	-2.96	-2.09	61.70	3.32	4.42	138.95
324	4.28	-8.51	-8.21	19.91	-12.21	-12.40	-12.05	39.56	-2.54	-1.53	61.33	3.71	4.77	141.54
326	3.91	-7.20	-6.95	19.39	-11.54	-11.73	-11.43	38.69	-1.55	-0.45	60.50	4.63	5.69	143.42
328	3.54	-5.67	-5.52	18.83	-10.80	-10.98	-10.74	37.60	0.01	0.98	59.18	6.02	7.27	144.59
330	3.18	-3.94	-3.89	18.28	-10.29	-10.20	-10.01	36.47	1.72	2.55	57.50	8.03	9.36	145.04
332	2.83	-1.97	-2.01	17.74	-9.56	-9.42	-9.28	35.35	2.91	4.17	55.78	10.38	11.62	144.78
334	2.49	0.40	0.19	17.25	-8.89	-8.67	-8.59	34.29	4.81	5.77	54.11	12.56	13.92	143.81
336	2.15	3.20	2.82	16.87	-7.81	-8.08	-8.03	33.38	6.09	7.20	52.61	14.43	16.07	142.12
338	1.87	6.20	5.65	18.20	-10.17	-10.22	-10.05	34.29	4.34	5.65	52.96	13.83	15.45	139.71
340	1.58	9.90	9.13	17.92	-9.88	-9.82	-9.68	35.21	2.97	4.14	53.44	13.02	14.64	136.60
342	1.31	14.62	13.54	17.55	-9.39	-9.27	-9.17	34.51	3.97	5.18	53.85	12.36	13.95	132.77
344	1.04	21.01	19.48	17.16	-8.83	-8.67	-8.62	33.69	5.09	6.46	52.68	13.95	15.62	128.22
346	0.77	30.37	28.17	16.76	-8.21	-8.04	-8.03	32.83	6.41	7.84	51.37	15.97	17.56	122.97
348	0.51	45.98	42.87	16.36	-7.64	-7.38	-7.41	31.97	7.86	9.29	50.03	18.22	19.63	117.00
350	0.25	82.55	76.97	15.96	-6.99	-6.70	-6.78	31.12	9.68	10.78	48.69	19.95	21.80	110.32
352	-0.02	-	-	15.58	-6.18	-6.00	-6.14	30.29	10.77	12.31	47.36	22.02	24.04	102.93
354	-0.28	-	-	15.20	-5.74	-5.30	-5.49	29.48	12.21	13.85	46.07	24.20	26.31	94.82
356	-0.54	-	-	14.83	-5.00	-4.60	-4.84	28.70	13.89	15.40	44.82	26.49	28.60	86.00
358	-0.79	-	-	14.49	-3.96	-3.91	-4.19	27.96	15.16	16.93	43.62	28.71	30.89	76.48
360	-1.04	-	-	14.16	-3.09	-3.24	-3.57	27.27	16.56	18.41	42.49	30.63	33.13	66.24
362	-1.29	-	-	13.86	-2.29	-2.60	-2.98	26.63	17.98	19.83	41.44	33.07	35.30	55.28
364	-1.54	-	-	13.60	-2.42	-2.04	-2.45	26.06	19.24	21.13	40.50	34.70	37.32	43.62
366	-1.78	-	-	13.43	-2.27	-1.67	-2.11	25.63	20.09	22.14	39.72	36.30	39.02	31.25
368	-2.00	-	-	14.73	-5.04	-4.56	-4.83	26.76	17.55	19.37	40.55	34.71	37.03	18.16
370	-2.23	-	-	14.63	-4.72	-4.39	-4.68	27.92	14.96	16.68	41.53	32.19	34.79	4.37
372	-2.45	-	-	14.46	-4.56	-4.06	-4.37	27.67	15.44	17.18	42.47	30.36	32.69	-10.14
374	-2.67	-	-	14.29	-3.93	-3.73	-4.07	27.34	16.08	17.86	42.06	30.93	33.49	-25.35
376	-2.88	-	-	14.23	-3.62	-3.62	-3.97	27.12	16.64	18.31	41.67	31.89	34.26	-41.28
378	-3.14	-	-	14.82	-5.29	-4.89	-5.17	27.64	15.94	17.08	42.01	30.85	33.43	-57.92
380	-3.41	-	-	14.59	-4.94	-4.45	-4.77	28.19	14.08	15.82	42.45	29.86	32.42	-75.26

Table 3.7: Same as in table 3.2 but for $Z = 140$ even-even isotopes.

Z=140	Proton decay			Alpha decay				$^{8}_{4}Be$ Decay			$^{12}_{6}C$ Decay			Fission
A	Q_p	$logT_p$	Ni.	Q_α	$logT_\alpha$	UDL	Royer	Q_{Be}	$logT_{Be}$	UDL	Q_C	$logT_C$	UDL	Xu
290	11.28	-18.65	-18.44	23.60	-15.16	-15.65	-14.99	45.89	-7.26	-6.92	70.80	-3.22	-2.67	-6.52
292	10.92	-18.42	-18.17	23.14	-14.70	-15.21	-14.58	44.96	-6.68	-6.00	69.39	-1.90	-1.35	9.03
294	10.56	-18.17	-17.88	22.69	-14.05	-14.75	-14.16	44.05	-5.61	-5.07	68.00	-0.62	-0.01	23.85
296	10.21	-17.90	-17.59	22.25	-13.41	-14.30	-13.75	43.18	-4.55	-4.14	66.63	0.68	1.34	37.95
298	9.87	-17.62	-17.28	21.83	-13.29	-13.86	-13.34	42.34	-3.86	-3.22	65.32	2.55	2.68	51.33
300	9.53	-17.33	-16.96	21.45	-11.95	-13.45	-12.96	41.57	-2.79	-2.36	64.09	3.20	3.98	63.99
302	9.20	-17.02	-16.63	21.14	-12.71	-13.11	-12.65	40.90	-2.43	-1.60	62.99	4.32	5.17	75.93
304	8.88	-16.71	-16.30	21.03	-12.69	-13.01	-12.56	40.50	-2.05	-1.16	62.19	5.50	6.02	87.16
306	8.56	-16.38	-15.95	21.50	-12.87	-13.61	-13.13	40.87	-2.52	-1.70	62.23	4.95	5.88	97.66
308	8.25	-16.03	-15.59	21.59	-13.19	-13.75	-13.27	41.45	-3.38	-2.48	62.66	4.33	5.26	107.45
310	7.94	-15.66	-15.21	21.52	-13.16	-13.70	-13.23	41.51	-3.15	-2.62	63.15	3.95	4.59	116.52
312	7.64	-15.27	-14.82	21.52	-13.14	-13.73	-13.27	41.49	-3.17	-2.65	63.24	3.45	4.40	124.88
314	7.38	-14.94	-14.47	21.88	-13.22	-14.19	-13.70	41.89	-4.15	-3.21	63.60	2.91	3.88	132.51
316	7.14	-14.59	-14.12	22.07	-13.63	-14.44	-13.95	42.39	-4.50	-3.87	64.12	2.36	3.18	139.43
318	6.89	-14.21	-13.74	22.25	-13.85	-14.68	-14.17	42.49	-4.98	-4.05	64.72	1.41	2.41	145.63
320	6.62	-13.77	-13.30	22.49	-14.11	-14.98	-14.46	42.69	-5.33	-4.36	64.85	2.11	2.17	151.12
322	6.31	-13.25	-12.77	22.53	-14.36	-15.06	-14.55	42.88	-5.57	-4.64	64.99	0.91	1.92	155.89
324	5.98	-12.62	-12.15	22.30	-13.80	-14.84	-14.34	42.71	-4.46	-4.50	64.94	0.97	1.88	159.94
326	5.62	-11.88	-11.44	21.89	-13.56	-14.40	-13.94	42.11	-4.86	-3.86	64.45	1.67	2.35	163.28
328	5.27	-11.07	-10.65	21.39	-13.01	-13.85	-13.43	41.21	-3.47	-2.82	63.47	2.83	3.39	165.90
330	4.91	-10.15	-9.76	20.86	-12.81	-13.23	-12.86	40.15	-2.58	-1.55	62.07	4.16	4.96	167.81
332	4.55	-9.14	-8.77	20.32	-12.22	-12.58	-12.25	39.05	-1.27	-0.16	60.44	5.82	6.87	169.00
334	4.19	-7.97	-7.66	19.79	-11.82	-11.91	-11.63	37.96	0.33	1.27	58.77	7.61	8.92	169.48
336	3.83	-6.63	-6.39	19.28	-11.19	-11.25	-11.01	36.92	1.50	2.70	57.15	9.60	11.01	169.25
338	3.47	-5.08	-4.92	18.87	-10.74	-10.69	-10.49	36.02	2.77	3.99	55.67	11.86	12.97	168.30
340	3.13	-3.39	-3.32	20.19	-11.99	-12.53	-12.23	36.96	1.46	2.53	56.09	11.12	12.30	166.63
342	2.79	-1.40	-1.43	19.82	-11.93	-12.08	-11.81	37.92	-0.08	1.10	56.62	10.23	11.46	164.26
344	2.45	1.03	0.81	19.36	-11.42	-11.48	-11.26	37.18	0.93	2.09	57.09	9.36	10.74	161.16
346	2.12	3.93	3.53	18.87	-10.79	-10.82	-10.64	36.32	2.08	3.29	55.90	11.09	12.31	157.36
348	1.80	7.49	6.92	18.38	-10.10	-10.13	-10.00	35.43	3.30	4.59	54.58	12.61	14.11	152.84
350	1.47	12.19	11.29	17.89	-9.45	-9.40	-9.33	34.54	4.60	5.94	53.24	14.33	16.03	147.61
352	1.16	18.61	17.23	17.41	-8.85	-8.67	-8.64	33.66	5.98	7.33	51.90	16.36	18.03	141.67
354	0.85	27.93	25.94	16.94	-8.22	-7.92	-7.95	32.80	7.52	8.75	50.58	18.85	20.06	135.02
356	0.56	43.60	40.51	16.50	-7.40	-7.19	-7.26	31.97	8.80	10.17	49.30	21.53	22.12	127.65
358	0.27	78.68	73.45	16.08	-6.80	-6.47	-6.59	31.18	9.97	11.58	48.08	22.36	24.16	119.57
360	0.00	-	-	15.70	-5.92	-5.78	-5.95	30.43	11.85	12.96	46.92	23.99	26.18	110.78
362	-0.27	-	-	15.35	-5.58	-5.13	-5.35	29.73	12.57	14.29	45.83	25.92	28.13	101.27
364	-0.52	-	-	15.03	-5.02	-4.53	-4.79	29.09	13.78	15.55	44.82	27.65	30.00	91.06
366	-0.77	-	-	14.76	-4.51	-4.00	-4.30	28.52	14.89	16.71	43.91	30.77	31.74	80.13
368	-1.02	-	-	14.58	-3.90	-3.65	-3.98	28.08	15.72	17.61	43.16	31.69	33.20	68.50
370	-1.26	-	-	15.89	-6.53	-6.28	-6.46	29.22	14.05	15.12	44.04	28.84	31.32	56.15
372	-1.49	-	-	15.76	-6.38	-6.08	-6.27	30.39	11.04	12.70	45.04	27.21	29.25	43.09
374	-1.72	-	-	15.57	-5.95	-5.74	-5.96	30.10	11.83	13.22	46.01	25.07	27.31	29.32
376	-1.96	-	-	15.38	-5.85	-5.39	-5.64	29.74	12.16	13.90	45.56	25.91	28.08	14.84
378	-2.19	-	-	15.28	-5.58	-5.23	-5.49	29.47	13.26	14.38	45.14	26.42	28.81	-0.35
380	-2.38	-	-	15.87	-6.76	-6.39	-6.59	30.00	11.68	13.26	45.51	25.69	28.02	-16.25

Table 3.8: Same as in table 3.2 but for $Z = 142$ even-even isotopes.

Z=142	Proton decay			Alpha decay				$^{8}_{4}Be$ Decay			$^{12}_{6}C$ Decay			Fission
A	Q_p	$logT_p$	Ni.	Q_α	$logT_\alpha$	UDL	Royer	Q_{Be}	$logT_{Be}$	UDL	Q_C	$logT_C$	UDL	Xu
290	12.18	-19.07	-18.97	24.49	-14.79	-16.14	-15.45	48.92	-9.75	-9.16	74.76	-5.15	-5.26	-13.54
292	11.81	-18.87	-18.72	24.04	-15.13	-15.71	-15.06	48.00	-8.56	-8.32	73.38	-4.50	-4.07	3.46
294	11.45	-18.66	-18.46	23.59	-14.72	-15.29	-14.66	47.09	-8.08	-7.47	72.00	-3.35	-2.83	19.75
296	11.10	-18.43	-18.19	23.15	-14.15	-14.86	-14.26	46.19	-7.05	-6.60	70.63	-2.13	-1.58	35.32
298	10.75	-18.19	-17.91	22.73	-13.58	-14.43	-13.87	45.32	-6.43	-5.73	69.30	-0.84	-0.32	50.16
300	10.41	-17.94	-17.63	22.34	-13.35	-14.03	-13.50	44.48	-5.58	-4.88	68.03	0.58	0.92	64.29
302	10.08	-17.67	-17.34	21.99	-13.15	-13.66	-13.16	43.72	-4.83	-4.07	66.84	1.48	2.10	77.70
304	9.76	-17.40	-17.04	21.71	-12.80	-13.36	-12.89	43.06	-4.03	-3.37	65.79	2.48	3.17	90.38
306	9.44	-17.12	-16.74	21.63	-11.72	-13.31	-12.85	42.67	-3.14	-2.98	65.05	3.12	3.91	102.35
308	9.13	-16.83	-16.43	22.12	-13.33	-13.92	-13.43	43.05	-4.26	-3.50	65.14	3.10	3.71	113.60
310	8.82	-16.51	-16.10	22.22	-13.37	-14.07	-13.57	43.62	-4.80	-4.23	65.61	2.43	3.08	124.13
312	8.52	-16.18	-15.75	22.16	-13.31	-14.03	-13.55	43.65	-5.18	-4.32	66.13	1.56	2.41	133.95
314	8.21	-15.85	-15.39	22.17	-13.33	-14.08	-13.60	43.59	-5.20	-4.32	66.20	1.98	2.23	143.04
316	7.96	-15.54	-15.08	22.64	-14.02	-14.64	-14.13	44.06	-5.62	-4.92	66.64	0.78	1.65	151.42
318	7.71	-15.22	-14.75	22.84	-14.30	-14.90	-14.38	44.62	-6.26	-5.62	67.25	0.63	0.89	159.08
320	7.46	-14.88	-14.40	22.99	-14.02	-15.10	-14.57	44.96	-6.98	-6.06	67.90	-0.56	0.10	166.03
322	7.18	-14.48	-13.99	23.16	-14.87	-15.32	-14.79	45.31	-7.30	-6.51	68.17	-1.04	-0.28	172.26
324	6.86	-13.99	-13.51	23.11	-14.62	-15.29	-14.77	45.50	-7.68	-6.78	68.32	-1.50	-0.54	177.77
326	6.52	-13.43	-12.94	22.81	-14.41	-14.99	-14.49	45.24	-7.22	-6.56	68.21	-1.49	-0.50	182.56
328	6.17	-12.79	-12.32	22.36	-13.16	-14.53	-14.06	44.56	-6.76	-5.87	67.59	-0.89	0.07	186.64
330	5.82	-12.09	-11.63	21.86	-13.11	-13.98	-13.56	43.65	-5.83	-4.90	66.49	1.37	1.17	190.01
332	5.47	-11.30	-10.87	21.33	-12.96	-13.38	-13.00	42.62	-4.50	-3.76	65.06	2.06	2.69	192.66
334	5.11	-10.45	-10.04	20.80	-12.22	-12.75	-12.42	41.56	-2.90	-2.53	63.47	3.33	4.44	194.59
336	4.76	-9.50	-9.12	20.28	-11.99	-12.12	-11.83	40.50	-2.27	-1.25	61.85	5.11	6.30	195.81
338	4.41	-8.44	-8.10	19.80	-11.47	-11.50	-11.26	39.49	-1.01	0.03	60.28	6.95	8.19	196.31
340	4.06	-7.21	-6.95	19.40	-10.69	-10.98	-10.78	38.59	0.04	1.20	58.84	8.62	9.97	196.10
342	3.72	-5.87	-5.67	20.77	-12.61	-12.84	-12.53	39.54	-1.22	-0.16	59.31	8.02	9.27	195.18
344	3.38	-4.33	-4.22	20.41	-11.69	-12.41	-12.13	40.50	-2.22	-1.48	59.89	7.28	8.43	193.54
346	3.05	-2.55	-2.54	19.96	-9.44	-11.85	-11.61	39.68	-0.94	-0.47	60.40	6.37	7.69	191.19
348	2.71	-0.51	-0.58	19.49	-10.69	-11.23	-11.03	38.75	-0.38	0.74	59.18	8.03	9.18	188.12
350	2.38	1.97	1.74	19.00	-10.31	-10.57	-10.42	37.77	0.84	2.05	57.84	9.45	10.89	184.35
352	2.06	5.00	4.55	18.52	-9.66	-9.88	-9.78	36.79	2.86	3.42	56.47	11.34	12.70	179.85
354	1.74	8.69	8.03	18.04	-8.27	-9.18	-9.13	35.83	3.64	4.83	55.10	13.03	14.58	174.65
356	1.43	13.44	12.48	17.58	-8.64	-8.48	-8.48	34.89	4.92	6.26	53.76	14.75	16.50	168.73
358	1.14	19.80	18.39	17.15	-8.11	-7.79	-7.84	33.99	6.31	7.69	52.47	16.61	18.42	162.10
360	0.85	28.69	26.71	16.75	-7.17	-7.13	-7.23	33.15	8.17	9.07	51.24	18.40	20.31	154.76
362	0.58	42.76	39.80	16.39	-6.82	-6.52	-6.66	32.37	8.83	10.39	50.09	20.11	22.14	146.71
364	0.32	70.27	65.60	16.06	-6.37	-5.95	-6.13	31.66	10.42	11.64	49.01	21.94	23.91	137.94
366	0.07	191.50	179.44	15.77	-5.88	-5.43	-5.64	31.02	11.16	12.81	48.02	23.39	25.58	128.46
368	-0.17	-	-	15.52	-5.37	-4.96	-5.22	30.45	12.14	13.86	47.13	25.31	27.14	118.27
370	-0.42	-	-	15.35	-5.10	-4.67	-4.94	30.01	12.99	14.68	46.39	26.92	28.44	107.37
372	-0.65	-	-	16.68	-7.55	-7.18	-7.31	31.16	10.73	12.36	47.28	24.41	26.69	95.76
374	-0.89	-	-	16.55	-7.06	-6.98	-7.13	32.34	10.13	10.12	48.29	22.61	24.78	83.44
376	-1.12	-	-	16.35	-7.05	-6.66	-6.83	32.01	9.04	10.65	49.26	21.35	22.99	70.40
378	-1.36	-	-	16.15	-6.67	-6.32	-6.52	31.62	9.70	11.32	48.77	21.60	23.76	56.66
380	-1.60	-	-	16.04	-6.56	-6.14	-6.36	31.31	10.52	11.85	48.29	22.37	24.53	42.20

Table 3.9: Same as in table 3.2 but for $Z = 144$ even-even isotopes.

Z=144	Proton decay			Alpha decay				$^{8}_{4}Be$ Decay			$^{12}_{6}C$ Decay			Fission
A	Q_p	$logT_p$	Ni.	Q_α	$logT_\alpha$	UDL	Royer	Q_{Be}	$logT_{Be}$	UDL	Q_C	$logT_C$	UDL	Xu
290	13.02	-19.37	-19.41	25.18	-15.09	-16.39	-15.69	50.48	-10.36	-9.84	78.41	-7.39	-7.38	-21.06
292	12.65	-19.22	-19.18	24.73	-15.45	-15.99	-15.32	49.58	-9.51	-9.06	77.08	-6.71	-6.31	-2.58
294	12.29	-19.05	-18.94	24.29	-14.36	-15.58	-14.94	48.68	-8.66	-8.25	75.72	-5.61	-5.19	15.17
296	11.93	-18.85	-18.69	23.85	-14.31	-15.17	-14.56	47.79	-7.97	-7.42	74.37	-4.40	-4.02	32.21
298	11.58	-18.64	-18.44	23.43	-13.63	-14.75	-14.18	46.91	-7.17	-6.58	73.03	-3.23	-2.84	48.52
300	11.24	-18.42	-18.18	23.02	-13.69	-14.35	-13.81	46.07	-6.03	-5.76	71.73	-2.16	-1.66	64.11
302	10.90	-18.19	-17.91	22.66	-12.40	-13.98	-13.46	45.29	-5.12	-4.97	70.49	-0.96	-0.50	78.98
304	10.58	-17.94	-17.65	22.33	-13.20	-13.65	-13.16	44.57	-4.89	-4.23	69.34	-0.03	0.59	93.13
306	10.26	-17.71	-17.37	22.09	-12.98	-13.39	-12.93	43.97	-4.17	-3.62	68.32	0.90	1.57	106.57
308	9.96	-17.45	-17.10	22.05	-12.95	-13.38	-12.92	43.65	-3.96	-3.31	67.62	1.71	2.23	119.28
310	9.65	-17.18	-16.80	22.55	-13.44	-13.99	-13.51	44.08	-4.65	-3.87	67.74	1.33	2.01	131.27
312	9.34	-16.90	-16.50	22.66	-13.35	-14.15	-13.66	44.68	-5.42	-4.62	68.23	0.91	1.39	142.55
314	9.04	-16.59	-16.18	22.60	-13.25	-14.12	-13.64	44.73	-4.34	-4.74	68.74	-0.03	0.74	153.10
316	8.73	-16.27	-15.85	22.62	-13.47	-14.18	-13.70	44.68	-5.32	-4.75	68.79	0.91	0.60	162.94
318	8.48	-16.01	-15.56	23.20	-13.68	-14.85	-14.34	45.27	-6.08	-5.48	69.30	-0.88	-0.05	172.06
320	8.23	-15.71	-15.26	23.40	-13.15	-15.11	-14.58	45.94	-7.15	-6.28	69.98	-1.70	-0.84	180.47
322	7.97	-15.40	-14.93	23.53	-14.36	-15.28	-14.75	46.27	-7.49	-6.70	70.67	-2.53	-1.65	188.16
324	7.68	-15.01	-14.54	23.63	-14.98	-15.42	-14.89	46.52	-7.74	-7.03	71.11	-2.45	-2.19	195.13
326	7.36	-14.57	-14.08	23.49	-14.86	-15.31	-14.79	46.56	-7.93	-7.13	71.32	-3.24	-2.50	201.38
328	7.02	-14.05	-13.57	23.14	-14.53	-14.95	-14.46	46.15	-7.53	-6.76	71.16	-3.29	-2.42	206.92
330	6.67	-13.49	-13.01	22.68	-13.85	-14.47	-14.02	45.38	-6.36	-5.98	70.43	-2.61	-1.78	211.74
332	6.32	-12.86	-12.39	22.17	-12.40	-13.93	-13.52	44.43	-5.93	-4.98	69.25	-1.62	-0.64	215.84
334	5.97	-12.18	-11.73	21.65	-13.00	-13.34	-12.97	43.41	-4.83	-3.86	67.82	-0.20	0.80	219.23
336	5.62	-11.44	-11.00	21.13	-12.47	-12.74	-12.41	42.37	-3.52	-2.66	66.28	1.80	2.42	221.91
338	5.28	-10.64	-10.22	20.63	-11.60	-12.13	-11.85	41.33	-2.31	-1.44	64.71	3.16	4.12	223.87
340	4.94	-9.75	-9.36	20.17	-11.23	-11.55	-11.31	40.35	-1.27	-0.23	63.19	4.74	5.84	225.11
342	4.61	-8.77	-8.43	19.81	-11.10	-11.09	-10.88	39.50	-0.26	0.85	61.81	6.69	7.45	225.64
344	4.26	-7.68	-7.36	21.19	-12.67	-12.94	-12.63	40.49	-1.65	-0.54	62.30	5.55	6.76	225.46
346	3.92	-6.42	-6.17	20.84	-12.26	-12.52	-12.24	41.51	-2.98	-1.90	62.89	4.87	5.94	224.56
348	3.59	-4.99	-4.81	20.39	-11.71	-11.97	-11.73	40.70	-2.04	-0.93	63.41	4.22	5.23	222.95
350	3.25	-3.35	-3.28	19.92	-11.22	-11.37	-11.17	39.78	-0.72	0.22	62.12	5.46	6.71	220.62
352	2.92	-1.49	-1.50	19.44	-10.67	-10.73	-10.58	38.82	0.43	1.48	60.70	7.32	8.42	217.58
354	2.60	0.74	0.58	18.96	-10.04	-10.06	-9.96	37.86	1.54	2.79	59.25	9.02	10.23	213.83
356	2.28	3.34	3.04	18.49	-9.02	-9.39	-9.33	36.90	2.88	4.15	57.80	10.58	12.12	209.36
358	1.96	6.48	5.99	18.04	-8.87	-8.72	-8.71	35.98	4.24	5.51	56.38	13.15	14.03	204.18
360	1.66	10.33	9.56	17.63	-8.37	-8.08	-8.12	35.11	5.44	6.84	55.03	14.23	15.92	198.29
362	1.38	15.02	13.92	17.26	-7.80	-7.50	-7.58	34.31	6.63	8.10	53.77	15.94	17.75	191.69
364	1.12	20.92	19.46	16.94	-7.12	-6.97	-7.09	33.59	7.98	9.28	52.60	17.62	19.50	184.37
366	0.87	28.99	26.98	16.65	-6.60	-6.48	-6.64	32.93	8.89	10.37	51.53	19.81	21.15	176.34
368	0.62	40.99	38.32	16.38	-6.34	-6.03	-6.22	32.34	9.78	11.39	50.56	20.61	22.69	167.60
370	0.38	63.16	58.92	16.15	-6.10	-5.63	-5.85	31.82	10.65	12.31	49.68	22.12	24.11	158.15
372	0.14	127.11	119.21	16.00	-5.35	-5.38	-5.62	31.41	11.56	13.01	48.96	23.30	25.29	147.98
374	-0.10	-	-	17.34	-8.11	-7.80	-7.90	32.59	9.14	10.77	49.87	21.56	23.63	137.11
376	-0.33	-	-	17.21	-7.58	-7.61	-7.73	33.78	7.09	8.60	50.88	19.75	21.82	125.52
378	-0.57	-	-	17.01	-7.59	-7.30	-7.44	33.46	7.61	9.11	51.86	18.12	20.12	113.22
380	-0.80	-	-	16.81	-7.38	-6.99	-7.15	33.06	8.60	9.75	51.34	18.87	20.89	100.22

Among all the said decay modes nuclei with least $T_{1/2}$ is dominant one. It is conceived from the calculated values of $T_{1/2}$ for different decay modes that it depends on the neutron number corresponding to a particular charge number and also the nuclei with $Z \leq 140$ which has a uniform decay mode with respect to the addition of neutron, i.e., these follow a sequence of proton, alpha, and SF successively. The region $Z \geq 140$ follows a different mode of decay sequence when the neutron is increased, and table 3.10 is self-explanatory in this regard. The half-life of light cluster decay (Be, C) is many orders higher than other decay modes, hence light cluster emission is unlikely to be observed in any of the isotopes of $Z = 130 - 144$ nuclei. Studies by Santhosh et al., (Santhosh and Sahadevan, 2010) with $N = Z$ (^{8}Be, ^{12}C, ^{16}O and others) and $N \neq Z$ cluster emission from isotopes of $^{132-138}$Nd, $^{144-158}$Gd, $^{176-196}$Hg and $^{192-198}$Pb suggest that $N = Z$ cluster emission is generally more probable as compared to $N \neq Z$ emission for small mass numbers, and for higher mass numbers $N \neq Z$ becomes equal or more probable (half-life few orders lower). Hence the salient features of dominant decay mode and trends in shell closure seen in the current study with N=Z light clusters will be identical for the decay of other $N \neq Z$ well-known clusters $^{10}_{4}Be$ and $^{14}_{6}C$.

Table 3.10: Prediction of dominant decay modes for $Z = 130 - 144$ isotopes

Parent	Region of dominating decay modes		
Z	1-Proton	α	SF
130	$290 \leq A \leq 298$	$300 \leq A \leq 342$	$344 \leq A \leq 380$
132	$290 \leq A \leq 304$	$306 \leq A \leq 348$	$350 \leq A \leq 380$
134	$290 \leq A \leq 306$	$308 \leq A \leq 356$	$358 \leq A \leq 380$
136	$290 \leq A \leq 308$	$310 \leq A \leq 362$	$364 \leq A \leq 380$
138	$290 \leq A \leq 314$	$316 \leq A \leq 370$	$372 \leq A \leq 380$
140	$290 \leq A \leq 318$	$320 \leq A \leq 378$	A=380
142	$290 \leq A \leq 320$	$322 \leq A \leq 380$	-
144	$292 \leq A \leq 324$	$326 \leq A \leq 380$	A=290

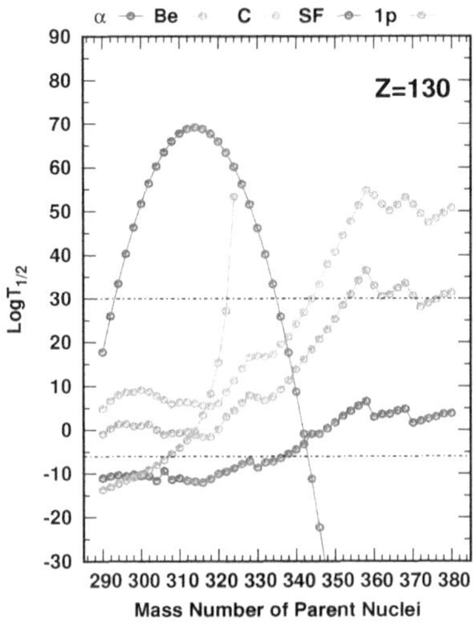

(a) Z = 130

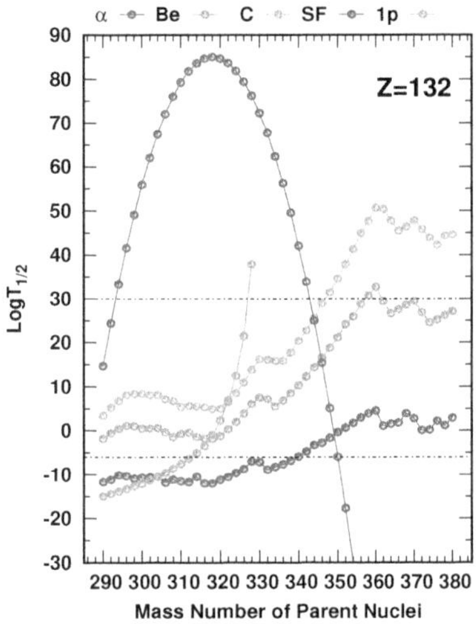

(b) Z = 132

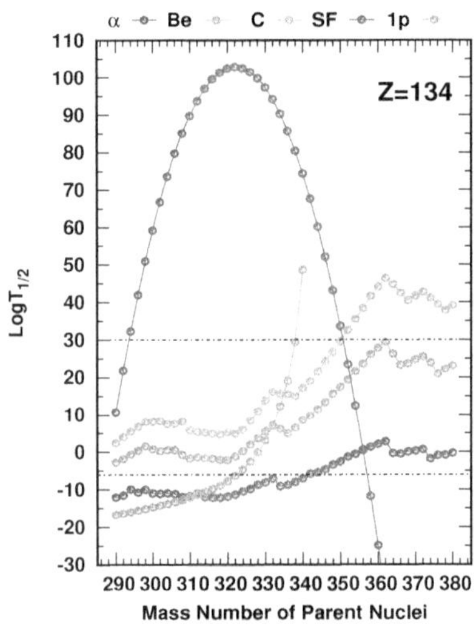

(c) Z = 134

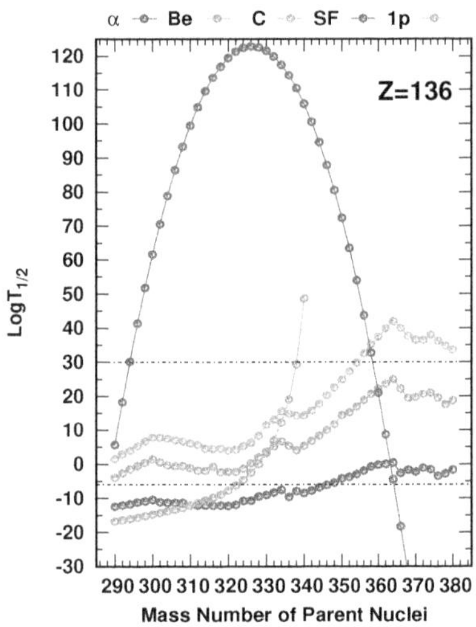

(d) Z = 136

Figure 3.11: Competition between the half life values against the 1proton emission, alpha decay, cluster decay and spontaneous fission for (a) Z = 130, (b) Z = 132, (c) Z = 134 and (d) Z = 136 isotopes. The curves for corresponding decay modes can be identified from the symbols given above the plots.

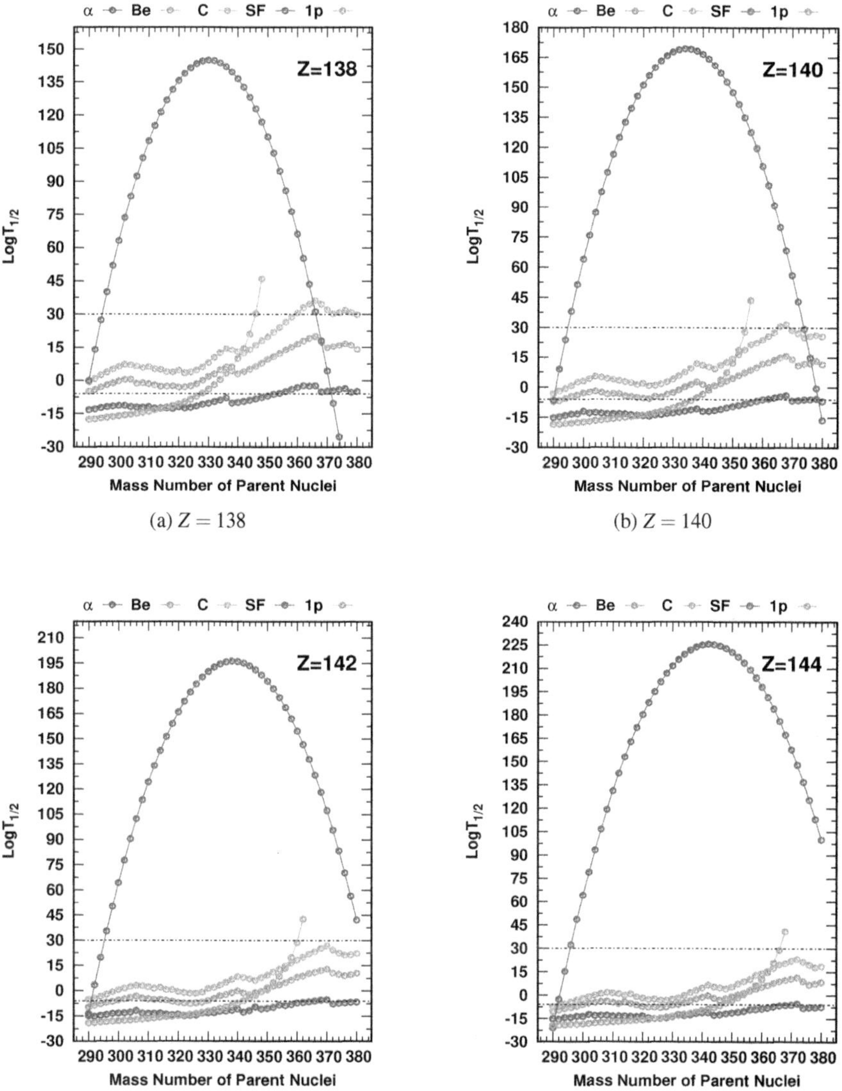

(a) Z = 138

(b) Z = 140

(c) Z = 142

(d) Z = 144

Figure 3.12: Same as in fig. 3.11 but for (a) Z = 138, (b) Z = 140, (c) Z = 142, (d) Z = 144 isotopes.

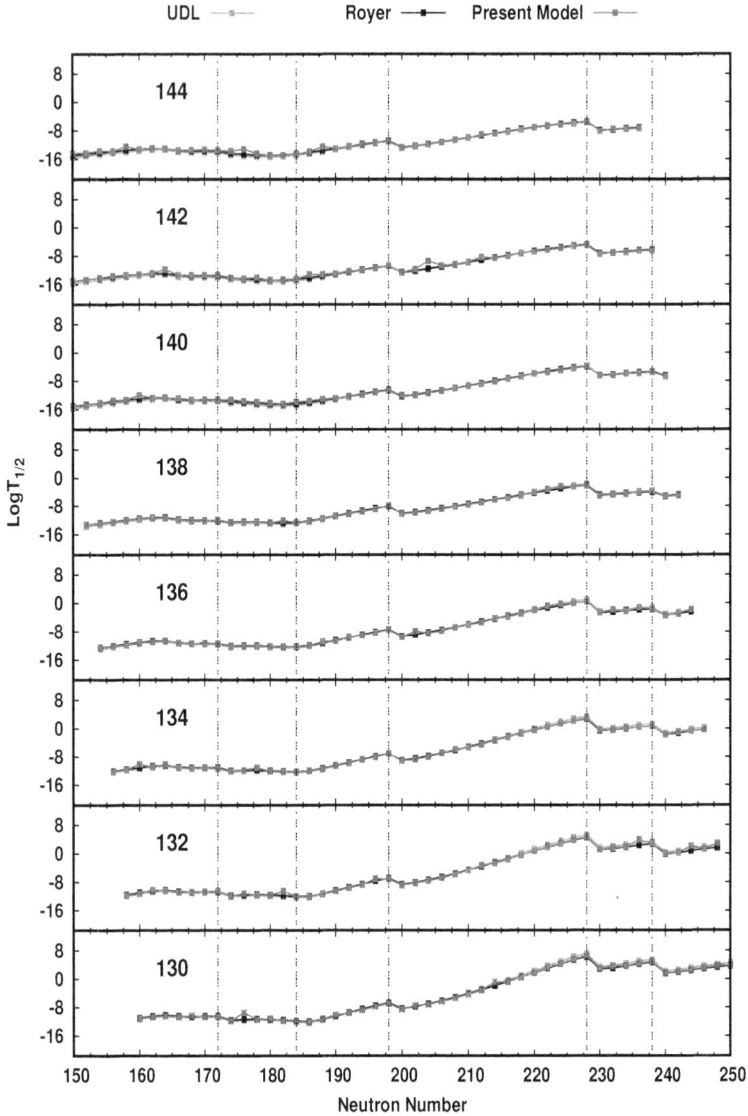

Figure 3.13: The alpha decay $logT_{1/2}$ half-lives by our model (Present model), Royer formula and UDL plotted against Parent nuclei neutron number. The vertical dotted lines indicate the predicted shell closures, where higher $T_{1/2}$ values are obtained. The numbers on the plot indicate corresponding Z values.

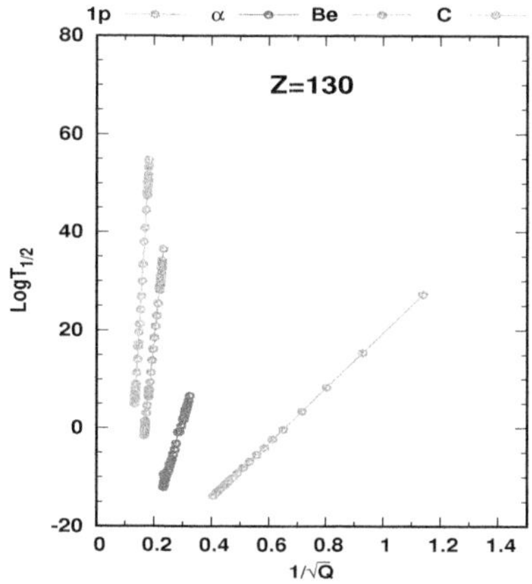

Figure 3.14: Geiger-Nuttall law curve for decay modes of $Z = 130$ isotope

The SHN when synthesized in the laboratory, its identification is done by the decay counts received which seems to be alpha emission or SF. The fig. 3.11 & 3.12 and table 3.10 indicate the regions which could be of experimental interest in order to decide the possible isotopes which can be synthesized based on SHN post-synthesis decay properties and half-life measurements. From table 3.2 - 3.9 it can be seen that proton decay, alpha decay, and cluster decay half-lives from the present model are in good agreement with other theoretical empirical formulae. In fig. 3.13 the alpha decay half-lives for $Z = 130 - 144$ isotopes calculated using the present model is compared with the values obtained from UDL (Qi et al., 2009) and Royer's formula (Royer, 2000).

The partial α-decay half-life is related to the energy of the escaping α-particle by the Geiger-Nuttal equation, which is derived by fitting experimental data for each isotopic chain. Accordingly, the equation is formulated in 1911 by Geiger and Nuttal as (Geiger and Nuttall, 1911),

$$log_{10}T_{1/2} = \frac{A(Z)}{Q_{\alpha}^{1/2}} + B(Z). \tag{3.81}$$

The Geiger-Nuttal law plot for decay modes of $Z = 130$ isotope is represented in fig. 3.14 as a sample plot. The slope and intercept differ from one mode to the other and the straight line indicates that the model predictions are correct. Still, the Geiger-Nuttal plots open up that the inclusion of proximity potential does not produce much deviation to the linear nature of this plot.

3.3 Alpha decay chain studies of superheavy elements

In the heavy and superheavy region, the dominant decay mode is either alpha decay or SF. Spontaneous fission in SHE occurs after a series of successive α-decay. The α-decay energy and half-life of these decay chains aid in comprehending nuclear structure studies of parent and daughter nuclei with the goal of providing information regarding islands of stability. In the superheavy region, the magic numbers are also an open question, as there is no consensus among different theoretical approaches, and experimental confirmation is much awaited. Till date, there has been several theoretical and experimental studies performed by various authors such as Oganessian et al., (Y. T. Oganessian, 2007; Y. T. Oganessian et al., 2006; Y. Oganessian et al., 2013; Utyonkov et al., 2017) and others to treasure out the islands of stability or the magic islands, aiming at synthesizing the SHN with long-lived alpha emitters beyond Z=114. Description of nuclear mass and binding energy provides the basic step in determining the stability of SHEs. To calculate the binding energy (B.E.) more precisely, macroscopic- microscopic models involving the sum of predominating macroscopic part (obtained from the liquid drop model of the atomic nucleus) and a microscopic part(from the nuclear shell model), are essential instead of just the liquid drop or the shell model (Hofmann et al., 2004). Based on these ideas, to calculate the binding energy, the simple Bethe-Weizsacker formula has been changed to account for the shell closure effects (Dong and Ren, 2005, 2008; Samanta and Adhikari, 2002). Similarly, the Finite-range liquid drop model (FRDM) calculations based on the finite-range droplet macroscopic and the folded-Yukawa single-particle microscopic nuclear-structure (Moller et al., 2016), and Skyrme energy-density functional theory based on macroscopic-microscopic mass Weizsacker-Skyrme (WS) formula (Wang et al., 2014) have been studied to give a good estimate to nuclear binding energy both theoretically and also in reproducing the experimental values. Furthermore, the mean-field models have also been successful in describing the atomic nuclei and their ground-state properties (Lalazissis et al., 2005; Nikšić et al., 2014). For the SHE, $Z = 122$ Weizsacker-Skyrme (WS4) mass model is used, and for $Z = 126$ & $Z = 138$, a DIRHB computer code with relativistic mean-field DDPC1 effective interaction (Nikšić et al., 2008; Agbemava et al., 2014; Nikšić et al., 2014) is used to obtain the ground-state properties of

Parent and daughter nuclei along with the WS4 mass model. The trends in the decay chain may provide signatures of subshell and shell closures in the superheavy region. The predictions in the present study may have an impact on the experimental synthesis and detection of new isotopes in near future. In this context, we focus on decay chains originating from $Z = 122$, $Z = 126$, and $Z = 138$ using the Cubic plus proximity potential model with an improved transfer matrix as discussed in chapter 2.

3.3.1 α-decay of $Z = 122$

A detailed study on the decay properties of $Z = 122$ isotopes which is likely to be synthesized in the near future, and in turn will give the experimentalist a chance to observe the decays associated with these particular isotope is discussed. To theoretically predict the features of alpha chain from Z=122 isotope in the mass range $298 \leq A \leq 350$ are studied using a fission-like model with an effective combination of the cubic plus proximity potential in the pre and post-scission regions, wherein the decay rates are calculated using improved transfer matrix method. To begin with the accuracy of alpha decay half life calculations with the experimental data is validated using a recently synthesized SHN and a comparative study was performed, which was found to be in good relevance with the current model (Cubic plus proximity potential with Improve transfer matrix method) as discussed in table 3.11. Decay half-life calculations are sensitive to choice of Q-value and there are several mass models for determining the Q-value. For every 1 MeV difference in Q-value, the decay half-life changes by several orders. As a result, the binding energy from the WS-4 model is used to calculate the Q-value of the decay process.

In decay chain, the first step is alpha decay originating from $Z = 122$ isotope and daughter nuclei created in this step acts as a parent for second step of the decay chain. This series of alpha emission is due to alpha decay half-life being several orders less than SF, implies that alpha decay is the dominant decay mode. In the subsequent steps when SF becomes the dominant decay mode, the decay chain ends. In table 3.12 & 3.13, the alpha decay half-life calculated from this model is compared with empirical formulas, and it is found that the model calculation is close to the values from different reliable empirical formula and matches well with UDL values. Further the modes of decay are given in the last column of the table. The alpha decay

Table 3.11: Comparison of theoretically calculated alpha decay half life of some SHN with the corresponding experimental values

Parent Nuclei	Q (MeV)		α- $log_{10}T_{1/2}$	
	Exp.	WS-4	Exp.	Current model
^{298}Og	11.820	12.202	-3.161	-4.512
^{293}Lv	10.710	10.797	-1.244	-1.672
^{292}Lv	10.780	11.130	-1.886	-2.135
^{291}Lv	10.890	11.124	-1.721	-2.348
^{290}Lv	11.000	11.088	-2.081	-2.454
^{290}Mc	10.410	10.287	-0.187	-0.495
^{289}Mc	10.490	10.299	-0.481	-0.517
^{288}Mc	10.630	10.401	-0.785	-0.979
^{287}Mc	10.760	10.505	-1.432	-1.089
^{284}Nh	10.120	10.119	-0.041	-0.034
^{283}Nh	10.380	10.412	-1.125	-0.731

half-life and SF half-life obtained for each subsequent step of decay chain originating from different isotopes is represented in figure 3.15 - 3.17. The crossing of SF curve with alpha decay curve indicates closure of alpha chain, and from this the chain length can be found. The isotopes of $Z = 122$ exhibit decay chains of varying lengths. The decay chains from isotopes $^{298-300,302,304-306,308-310,312,314}122$ is expected to show 7α, 5α, 4α, 3α, 2α and 1α decay respectively. For decay chains originating from isotopes with A = 298 to 316, the length of chain is seen to decrease with increase in mass number. All the other isotopes from A = 316 to 350 may undergo SF since the obtained SF half-lives are comparatively less. The Q-values and half-lives associated are unique for each chain. In decay chain, the nuclei produced in subsequent steps have different Z and A values. Some of the nuclei produced are close to neutron magic numbers. The variation of alpha decay curve in decay chain, exhibits next minimum in the half-life curves at N = 186, 178 & 164. This could be seen as rapid decay of nuclei at this step to attain closed-shell configuration, which suggest the subshell or shell closure at N = 184, 176 & 162. Among these N=184 is a magic number which is a common predication from different formalism. Also the two closely lying magic numbers 176 and 184, agrees well with the predictions of two-center shell model approach (R. Gherghescu et al., 1998; R. A. Gherghescu and Greiner, 2012). It is to be noted that, theoretical decay chains are indicating that magicity imprint would be present in the decay chain, and the magicity in superheavy region could be established if similar signatures could be seen in experimentally detected decay chains.

Table 3.12: The Q_α, $T_{1/2}$ and decay mode of isotopes derived in the α-decay chain of $^{298,300,302,304,306}122$

Parent Nuclei	Q (MeV)	$log_{10}T_{1/2}$ (alpha decay)				$log_{10}T_{1/2}$ (fission)	Decay Mode
		Model	UDL	Royer	VS		
$^{298}122$	14.7075	-7.9557	-8.2394	-7.6465	-8.0399	23.5852	α
$^{294}120$	13.2460	-5.8002	-5.9340	-5.4753	-5.8532	16.2047	α
$^{290}118$	12.6049	-5.0047	-5.1357	-4.7113	-5.0840	10.2483	α
$^{286}116$	11.3164	-2.5677	-2.6827	-2.4106	-2.7551	5.6268	α
$^{282}114$	11.3817	-3.1524	-3.4930	-3.1498	-3.4996	2.2529	α
$^{278}112$	11.7833	-4.8373	-5.0943	-4.6265	-4.9903	0.0404	α
$^{274}110$	10.8721	-3.2913	-3.5025	-3.1374	-3.4701	-1.0954	α
$^{270}108$	9.0297	1.5517	1.1945	1.2319	0.9820	-1.2378	SF
$^{300}122$	14.2260	-7.3318	-7.3716	-6.8081	-7.2301	23.1121	α
$^{296}120$	13.3474	-4.9372	-6.1774	-5.6721	-6.0893	15.7046	α
$^{292}118$	12.2446	-4.6066	-4.3499	-3.9499	-4.3510	9.7209	α
$^{288}116$	11.2943	-2.9564	-2.6580	-2.3578	-2.7390	5.0722	α
$^{284}114$	10.5769	-1.6716	-1.4007	-1.1739	-1.5350	1.6708	α
$^{280}112$	10.8669	-2.8941	-2.8586	-2.5180	-2.8896	-0.5693	α
$^{276}110$	10.8887	-3.8287	-3.5775	-3.1772	-3.5482	-1.7328	α
$^{272}108$	9.5314	0.7737	-0.4493	-0.2651	-0.5767	-1.9031	SF
$^{302}122$	14.2409	-7.0626	-7.4320	-6.8346	-7.2937	21.9250	α
$^{298}120$	13.0112	-5.6464	-5.4987	-5.0104	-5.4573	14.4905	α
$^{294}118$	12.2026	-4.5167	-4.2842	-3.8590	-4.2961	8.4797	α
$^{290}116$	11.0888	-2.4550	-2.1538	-1.8590	-2.2712	3.8037	α
$^{286}112$	9.9739	0.0552	0.3248	0.4605	0.0838	0.3750	α
$^{282}110$	10.1446	-1.1714	-0.8903	-0.6585	-1.0412	-1.8927	SF
$^{304}122$	13.7424	-6.4920	-6.4839	-5.9219	-6.4084	20.0243	α
$^{300}120$	13.3229	-6.1920	-6.1904	-5.6249	-6.1152	12.5628	α
$^{296}118$	11.7561	-3.5338	-3.2451	-2.8623	-3.3247	6.5249	α
$^{292}116$	11.1308	-2.1354	-2.2963	-1.9621	-2.4125	1.8216	α
$^{288}114$	9.6497	1.0657	1.3056	1.4021	1.0008	-1.6345	SF
$^{306}120$	12.8940	-5.5930	-5.3078	-4.7737	-5.2913	9.9216	α
$^{302}118$	12.1867	-4.5959	-4.3104	-3.8245	-4.3347	3.8567	α
$^{298}116$	10.6687	-1.4170	-1.0729	-0.7945	-1.2671	-0.8738	α
$^{294}114$	9.5244	1.4350	1.6795	1.7790	1.3459	-4.3573	SF

Table 3.13: The Q_α, $T_{1/2}$ and decay mode of isotopes derived in the α-decay chain of $^{308,310,312,314,316}122$

Parent Nuclei	Q (MeV)	$log_{10}T_{1/2}$ (alpha decay)				$log_{10}T_{1/2}$ (fission)	Decay Mode
		Model	UDL	Royer	VS		
$^{308}122$	14.94453	-8.9093	-8.8257	-8.0444	-8.6249	14.0826	α
$^{304}120$	12.76728	-5.1563	-5.0604	-4.5140	-5.0652	6.5672	α
$^{300}118$	11.95991	-4.0984	-3.8044	-3.3243	-3.8650	0.4753	α
$^{296}116$	10.89668	-2.0150	-1.7334	-1.3799	-1.8963	-4.2823	SF
$^{310}122$	13.4606	-6.234	-6.000	-5.384	-5.974	10.042	α
$^{306}120$	13.7917	-7.402	-7.235	-6.509	-7.118	2.500	α
$^{302}118$	12.0449	-3.976	-4.039	-3.513	-4.092	-3.619	α
$^{298}116$	10.7744	-1.463	-1.430	-1.068	-1.617	-8.404	SF
$^{312}122$	12.1665	-3.4991	-3.1243	-2.6762	-3.2745	5.2889	α
$^{308}120$	12.9704	-5.6640	-5.5679	-4.9284	-5.5561	-2.2801	α
$^{304}118$	13.1264	-6.6791	-6.4810	-5.7571	-6.3984	-8.4260	SF
$^{314}122$	12.1208	-3.4124	-3.0441	-2.5728	-3.2058	-0.1769	α
$^{310}120$	11.5029	-2.4905	-2.1231	-1.6927	-2.3211	-7.7727	SF
$^{316}122$	11.6628	-2.2120	-1.9255	-1.5025	-2.1602	-6.3550	SF

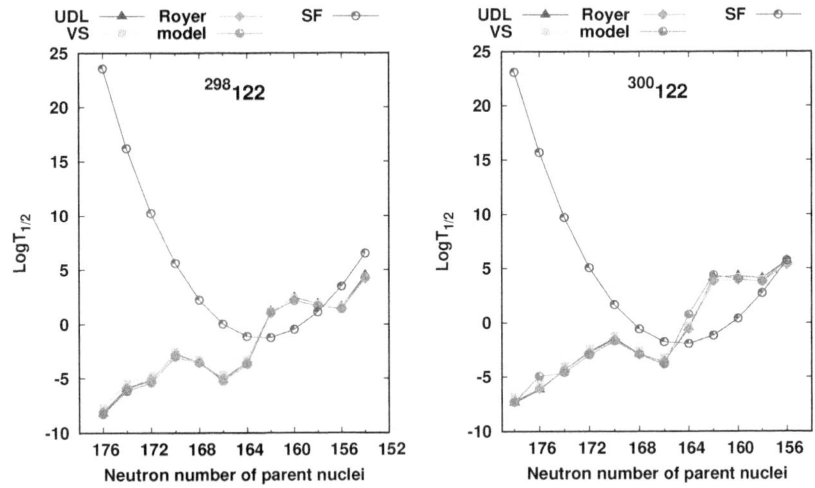

(a) $^{298}122$ (b) $^{300}122$

Figure 3.15: Decay chains originating from (a)$^{298}122$ & (b)$^{300}122$ isotopes.

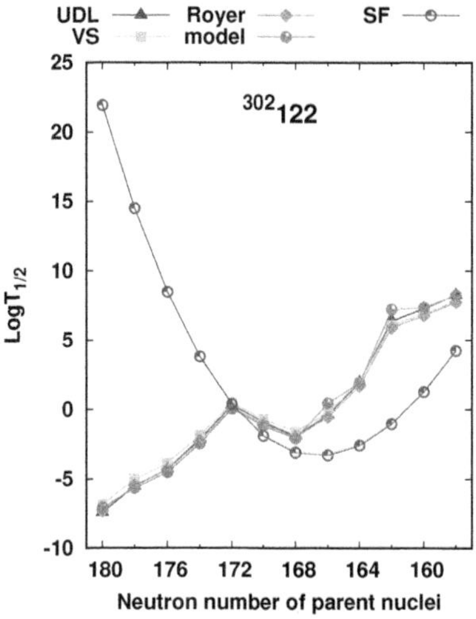

(a) $^{302}122$

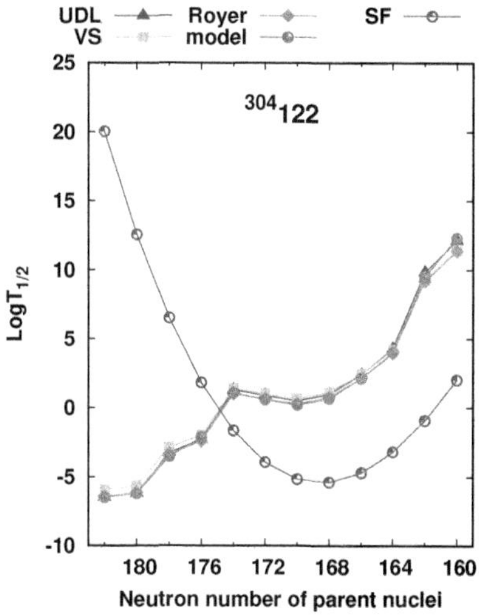

(b) $^{304}122$

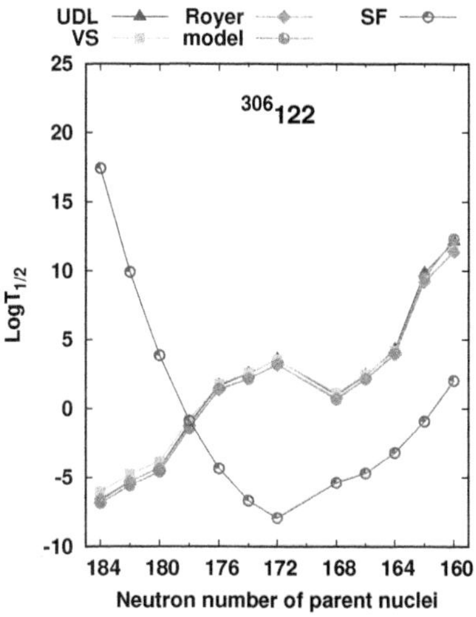

(c) $^{306}122$

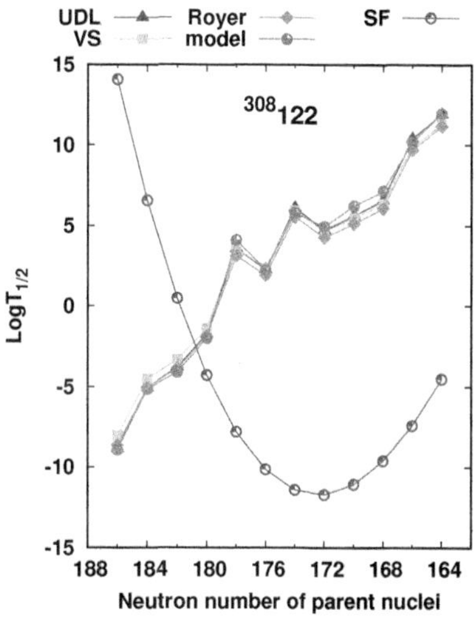

(d) $^{308}122$

Figure 3.16: Decay chains originating from (a)$^{302}122$, (b)$^{304}122$, (c)$^{306}122$ & (d)$^{308}122$ isotopes.

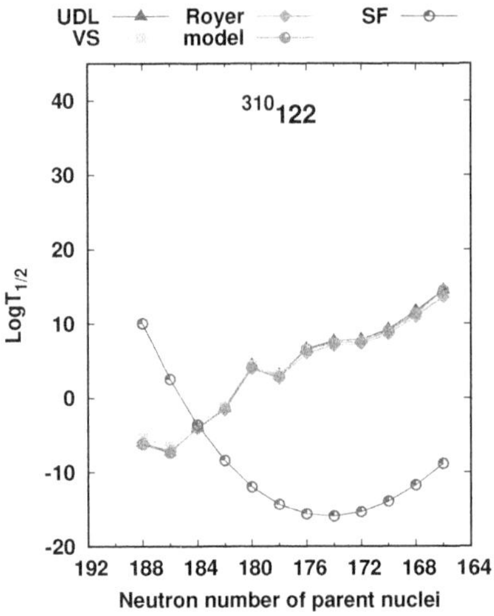

(a) $^{310}122$

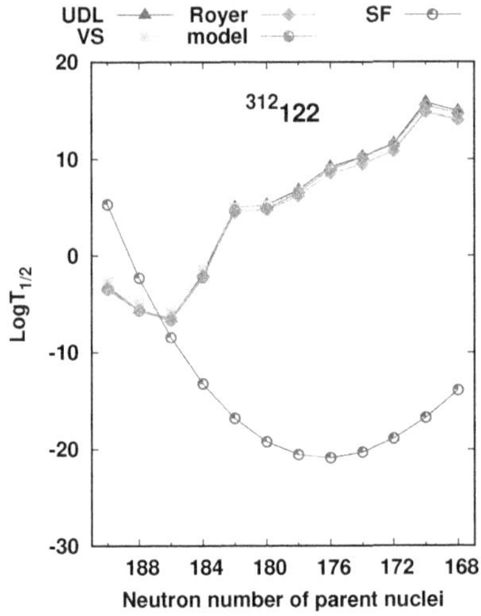

(b) $^{312}122$

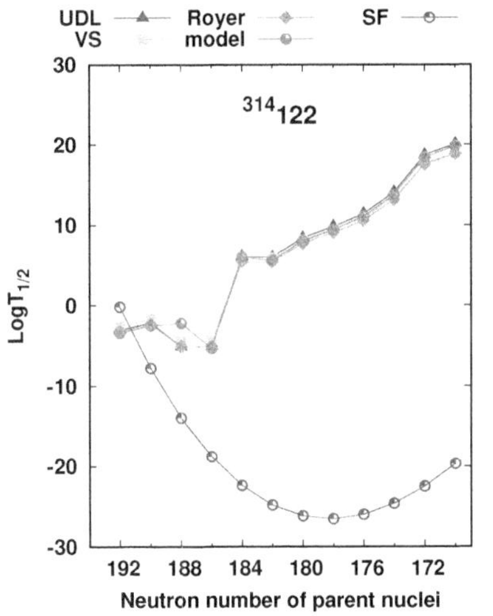

(c) $^{314}122$

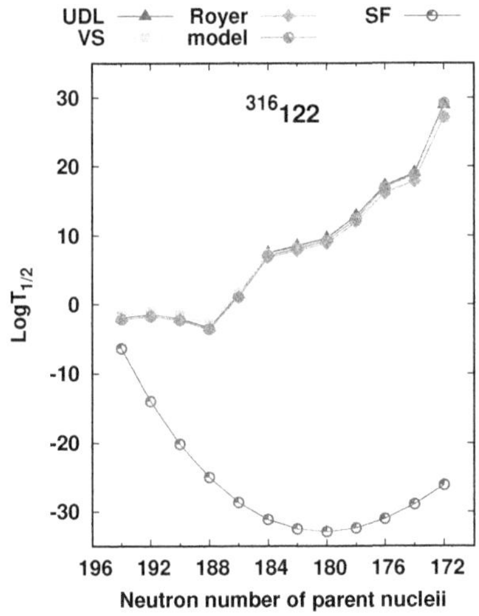

(d) $^{316}122$

Figure 3.17: Decay chains originating from (a)$^{310}122$, (b)$^{312}122$, (c)$^{314}122$ & (d)$^{316}122$ isotopes.

3.3.2 α-decay of $Z = 126$ & $Z = 138$

Here, the decay chains originating from the proton shell closures $Z = 126$ & $Z = 138$ is presented. The ground state properties of daughter and fragment are obtained from the DIRHB code with RMF DDPC1 effective interaction. For the Binding energy using WS (Weizsacker-Skyrme) model, the WS3.3 computer code (Wang et al., 2010) is executed since the WS4 B.E. data table is only up to $Z = 132$. The Q-values of $Z = 126$ isotopes using binding energy data from DDPC1, WS4, and FRDM are plotted in fig. 3.18 and it is found that the WS4 and DD- PC1 Q-values are close to each other.

The α decay half-lives calculated using the present model are analyzed with other mathematical / phenomenological formulae namely Universal decay law (UDL) of Qi et al., (Qi et al., 2009), the analytical formula of Royer (Royer, 2000), and Viola-Seaborg semi-empirical formula (VSS) (Viola and Seaborg, 1966). To study the SF half-lives of SHN, a parabolic potential approximation method with a semi-empirical formula by Xu et al., (Xu et al., 2008; Xu and Ren, 2005) is taken into consideration . The α-decay study of $Z = 126$ $(314 \leq A \leq 340)$ and $Z = 138$ $(350 \leq A \leq 400)$

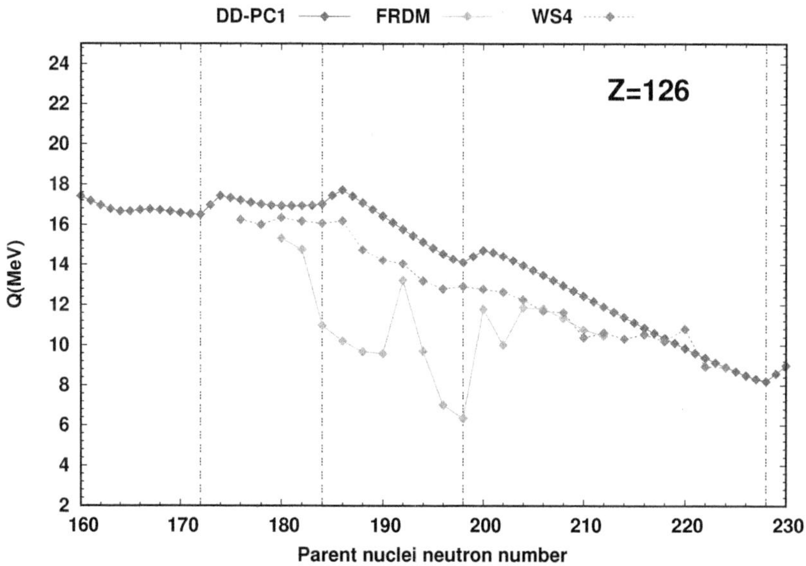

Figure 3.18: Comparison of alpha decay Q-values for $Z = 126$ isotopes obtained from DD-PC1 parametrization, FRDM and WS4 mass models.

are carried out using Cubic plus proximity potential with improved transfer matrix method. In the studied mass region 1p, 2p, 1n, 2n separation energies indicate a positive value and hence the α-decay via emission of 1-p, 2-p, 1-n, 2-n is prohibited, so alpha decay and SF are the only allowed decay modes. The competition between alpha and SF determines the decay mode associated with the isotope. Alpha decay chains from Z = 126 and Z = 138 isotope will constitute a series of alpha emissions where daughter nuclei formed in every step acts as parent nuclei for the next step of alpha emission. For the alpha decay chain to occur the first emission from the originating isotope must be an alpha particle; SF occurring at the very first step doesn't produce any alpha decay chain.

The parent nuclei of Z = 126 and Z = 138 and the subsequent parents formed through alpha emission is listed in column 1 of Table 3.14 - 3.18. The Q-values chosen for the calculation of alpha decay half-life using the present method with different mass models such as DD-PC1, WS4 (Z=126) and WS3.3 (Z=138) are given in column 2 and are found to be close to each other, and their corresponding alpha decay half-lives obtained by the present model (Column 3) is in good agreement with the half-lives obtained by their phenomenological formalism such as UDL, Royer, and VS given in column 4, 5 and 6. The SF calculated using Xu et al., (Xu et al., 2008; Xu and Ren, 2005) is given in column 7. The alpha decay chain originating from Z=126 and Z=138 are plotted with other formalisms; the UDL, Royer, and VS along with SF by Xu et al., in fig. 3.19 - 3.21. The intersection of the SF half-life curve with the alpha decay half-life curve indicates the termination of the alpha chain. The plots for Z = 126 indicate that long-chains originate from isotopes with small mass numbers, wherein chain originating from $^{314}126$ would emit 6α, while $^{316,318}126$ emit 5α, and from $^{320}126$ 4α particles. Smaller chains are obtained for $^{322,324,326,328}126$ with $3,3,2,1\alpha$ emission respectively. The isotopes of $Z = 126$ with $A \geq 330$ are likely to undergo SF and would not exhibit the alpha decay chain. For Z=138, the decay chains are listed in table 3.16,3.17, 3.18 whereas only longer chains are shown in figure 3.21. The decay chains associated with $^{350,352,354,356\&358}138$ consist of $8,7,6,5$ alpha emissions respectively, whereas the decay chains from $^{360,362\&364,366,368\&370}138$ consists of $4,3,2,1$ alpha emissions; hence it is clear that the decay chain length decreases with increase in mass number. At $A \geq 372$, $Z = 138$ undergoes SF.

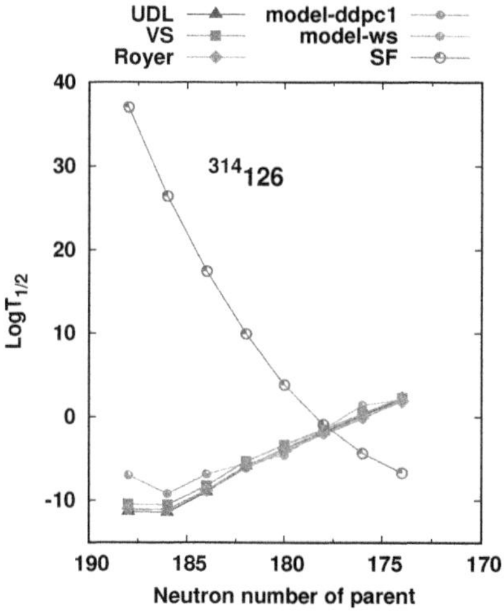

(a) $^{314}126$

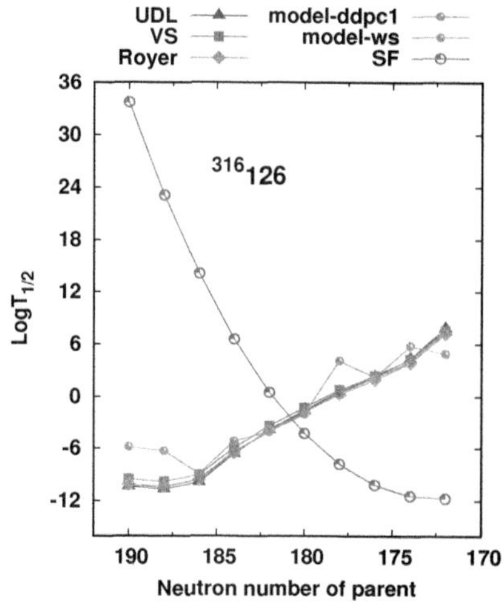

(b) $^{316}126$

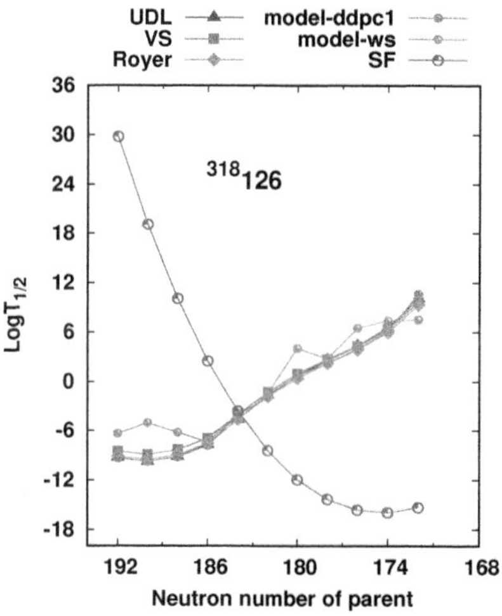

(c) $^{318}126$

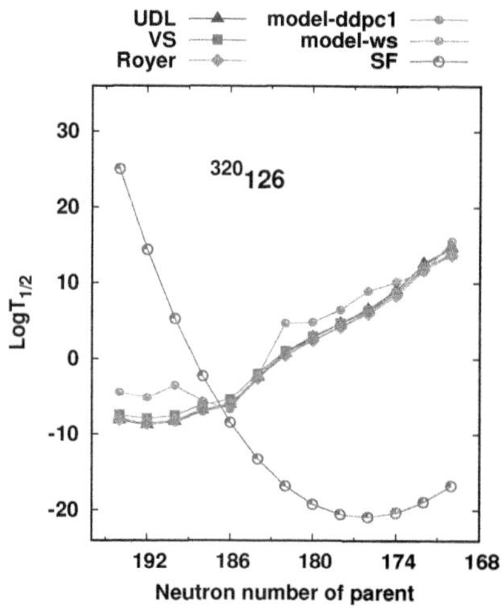

(d) $^{320}126$

Figure 3.19: Decay chains originating from (a)$^{314}126$, (b)$^{316}126$, (c)$^{318}126$ & (d)$^{320}126$ isotopes.

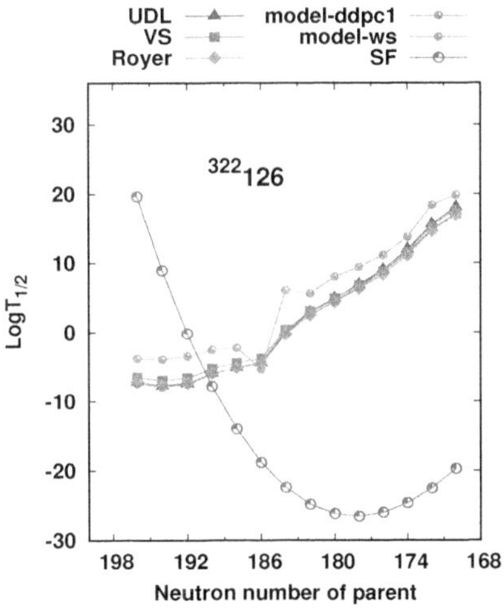

(a) ^{322}126

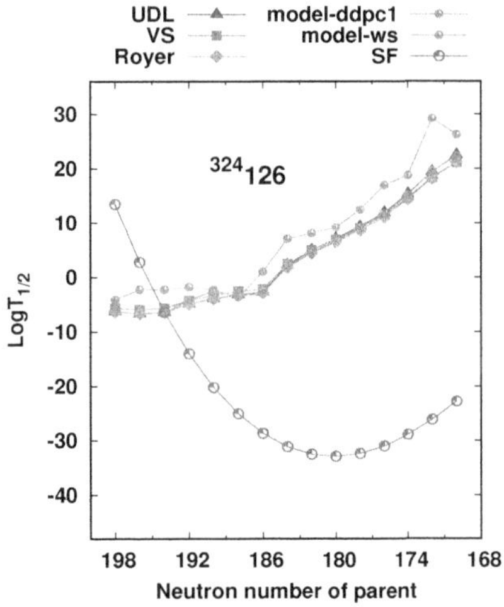

(b) ^{324}126

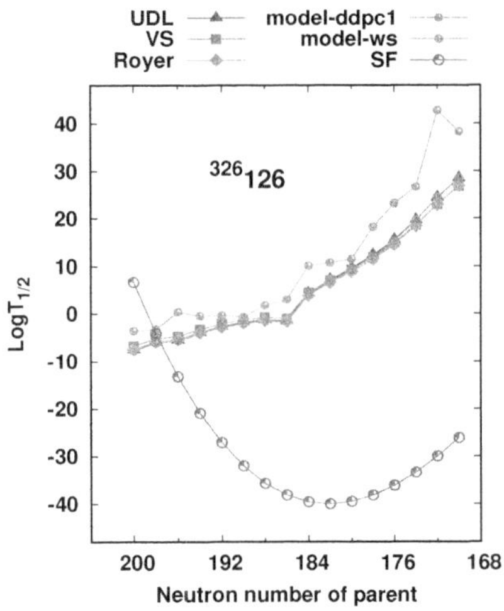

(c) ^{326}126

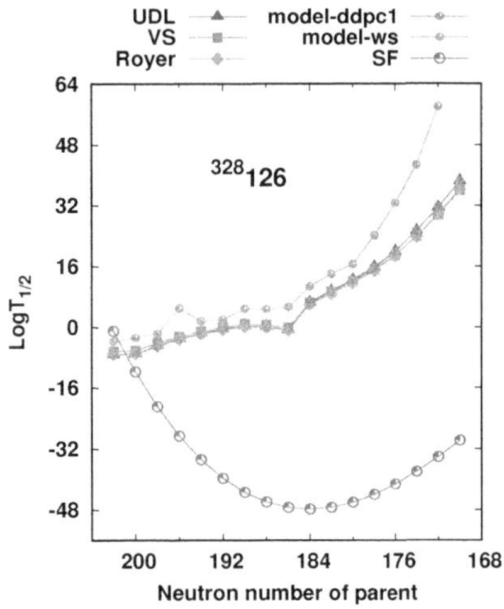

(d) ^{328}126

Figure 3.20: Same as in 3.19 for (a)^{322}126, (b)^{324}126, (c)^{326}126 & (d)^{328}126 isotopes.

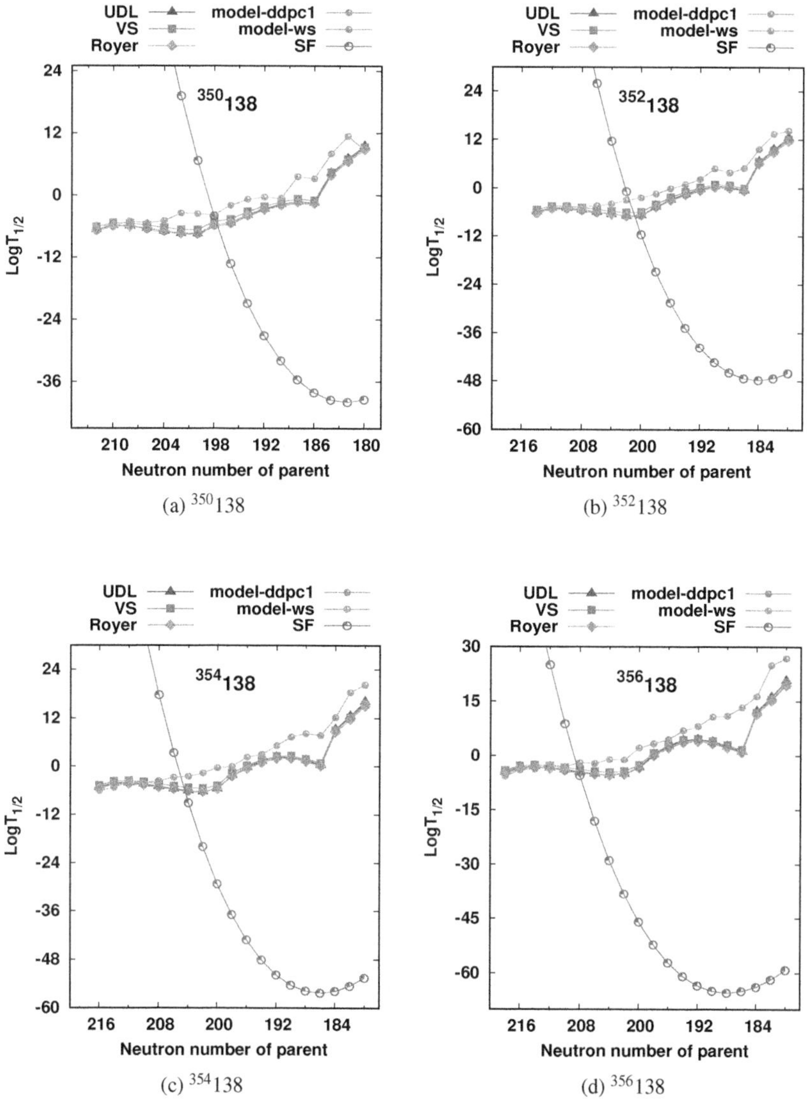

Figure 3.21: Decay chains originating from (a)^{350}138, (b)^{352}138, (c)^{354}138 & (d)^{356}138 isotopes.

Table 3.14: The Q_α, $T_{1/2}$ and decay mode of isotopes derived in the α-decay chain of $^{314}126$, $^{316}126$ and $^{318}126$

Parent Nuclei	Q (MeV)	$log_{10}T_{1/2}$ (alpha decay)				$log_{10}T_{1/2}$ (fission)	Decay Mode
		Model	UDL	Royer	VS		
$^{314}126$	17.093 [a]	-11.1394	-11.3188	-10.4538	-10.9940	37.0204	α
	14.756 [b]	-6.9756	-7.3977	-6.7989	-7.3076		
$^{310}124$	16.857 [a]	-11.0703	-11.4331	-10.5302	-11.0865	26.4123	α
	15.438 [b]	-9.1998	-9.1447	-8.3979	-8.9343		
$^{306}122$	15.068 [a]	-8.9157	-9.0127	-8.2479	-8.7943	17.4100	α
	13.807 [b]	-6.8690	-6.6461	-6.0435	-6.5677		
$^{302}120$	13.170 [a]	-6.1366	-5.9019	-5.3268	-5.8504	9.9215	α
	12.893 [b]	-5.5929	-5.3078	-4.7736	-5.2912		
$^{298}118$	11.974[a]	-4.1174	-3.8087	-3.3575	-3.8622	3.8566	α
	12.186 [b]	-4.5958	-4.3104	-3.8245	-4.3346		
$^{294}116$	10.923[a]	-2.0592	-1.7735	-1.4463	-1.9270	-0.8737	α
	10.668 [b]	-1.4170	-1.0728	-0.7945	-1.2671		
$^{290}114$	9.989[a]	-0.1675	0.2252	0.4266	-0.0243	-4.3572	SF
	9.523 [b]	1.4349	1.6794	1.7790	1.3458		
$^{316}126$	16.434[a]	-10.3056	-10.3306	-9.5033	-9.5128	33.7463	α
	14.237 [b]	-5.7828	-6.4292	-5.8669	-6.4036		
$^{312}124$	16.342 [a]	-10.3619	-10.6683	-9.7882	-10.3737	23.1116	α
	13.859 [b]	-6.2958	-6.2244	-5.6476	-6.1944		
$^{308}122$	15.541 [a]	-9.5178	-9.8577	-9.0055	-9.5957	14.0825	α
	14.944 [b]	-8.9093	-8.8257	-8.0443	-8.6248		
$^{304}120$	13.495 [a]	-6.3680	-6.6085	-5.9554	-6.5221	6.5672	α
	12.767 [b]	-5.1563	-5.0604	-4.5140	-5.0651		
$^{300}118$	11.978 [a]	-4.0995	-3.8488	-3.3657	-3.9069	0.4752	α
	11.959 [b]	-4.0984	-3.8043	-3.3242	-3.8650		
$^{296}116$	10.859 [a]	-1.7434	-1.6310	-1.2846	-1.7998	-4.2823	SF
	10.896 [b]	-2.0150	-1.7333	-1.3798	-1.8962		
$^{318}126$	15.778[a]	-9.3615	-9.2815	-8.4963	-9.0916	29.7592	α
	14.054 [b]	-6.3654	-6.0945	-5.5259	-6.0955		
$^{314}124$	15.758 [a]	-9.7507	-9.4297	-8.9026	-9.5158	19.0979	α
	13.241 [b]	-5.0643	-4.9754	-4.4548	-5.0264		
$^{310}122$	15.122 [a]	-9.2073	-9.1703	-8.3360	-8.9556	10.0421	α
	13.460 [b]	-6.2342	-6.0004	-5.3837	-5.9735		
$^{306}120$	14.019 [a]	-7.8100	-7.6800	-6.9238	-7.5371	2.4999	α
	13.798 [b]	-7.4016	-7.2347	-6.5093	-7.1181		
$^{302}118$	12.287 [b]	-4.0664	-4.6078	-4.0429	-4.6283	-3.6190	α
	12.044[a]	-3.9759	-4.0387	-3.5133	-4.0924		
$^{298}116$	10.852 [a]	-1.8596	-1.6448	-1.2685	-1.8198	-8.4037	SF
	10.774 [b]	-1.4627	-1.4295	-1.0682	-1.6170		

[a] Q-value calculated using binding energy obtained from DIRHBS code(with DD-PC1)
[b] Q-value calculated using binding energy obtained from the WS4 mass table.

The dips present in the alpha decay half-life curves at N=174, 186, 200 may be the signature of neutron shell closure at N=172, 184, and 198, which are inherently present in the WS and

Table 3.15: Same as in Table 3.14 for $^{320-328}126$

Parent Nuclei	Q (MeV)	$log_{10}T_{1/2}$ (alpha decay)				$log_{10}T_{1/2}$ (fission)	Decay Mode
		Model	UDL	Royer	VS		
$^{320}126$	15.140[a]	-7.8556	-8.1974	-7.4569	-8.0788	25.0593	α
	13.191[b]	-4.4461	-4.2985	-3.8231	-4.4136		
$^{316}124$	15.155[a]	-8.7216	-8.7440	-7.9370	-8.5769	14.3713	α
	13.199[b]	-5.1220	-4.9002	-4.3558	-4.9623		
$^{312}122$	14.622[a]	-8.3826	-8.3054	-7.5015	-8.1484	5.2888	α
	12.166[b]	-3.4991	-3.1243	-2.6761	-3.2745		
$^{308}120$	13.592[a]	-7.0281	-6.8677	-6.1385	-6.7793	-2.2801	α
	12.970[b]	-5.6639	-5.5678	-4.9283	-5.5560		
$^{304}118$	12.920[a]	-6.2555	-6.0455	-5.3519	-5.9884	-8.4260	SF
	13.126[b]	-6.6791	-6.4809	-5.7571	-6.3983		
$^{322}126$	14.548[a]	-7.3504	-7.1265	-6.4301	-7.0786	19.6467	α
	12.803[b]	-3.7492	-3.4506	-3.0042	-3.6231		
$^{318}124$	14.558[a]	-7.8534	-7.6840	-6.9206	-7.5866	8.9322	α
	12.568[b]	-3.8864	-3.5032	-3.0256	-3.6552		
$^{314}122$	14.107[a]	-7.4537	-7.3634	-6.5953	-7.2687	-0.1769	α
	12.120[b]	-3.4123	-3.0441	-2.5728	-3.2057		
$^{310}120$	13.106[a]	-5.9787	-5.8915	-5.2008	-5.8672	-7.7726	SF
	11.502[b]	-2.490	-2.1231	-1.6927	-2.3211		
$^{324}126$	14.109[a]	-6.1752	-6.2976	-5.6289	-6.3058	13.5217	α
	12.914[b]	-4.0818	-3.7368	-3.2424	-3.8987		
$^{320}124$	13.996[a]	-6.6980	-6.6249	-5.9052	-6.5971	2.7807	α
	11.913[b]	-2.2192	-1.9334	-1.5346	-2.1856		
$^{316}122$	13.595[a]	-6.5179	-6.3723	-5.6436	-6.3430	-6.3550	α
	11.662[b]	-2.2119	-1.9255	-1.5025	-2.1601		
$^{312}120$	12.623[a]	-4.0918	-4.8629	-4.2146	-4.9059	-13.9775	SF
	11.220[b]	-1.7590	-1.4075	-0.9981	-1.6545		
$^{326}126$	14.716[a]	-7.7204	-7.5084	-6.7287	-7.4502	6.6844	α
	12.784[b]	-3.5465	-3.1855	-3.4674	-2.9630		
$^{322}124$	13.595[a]	-6.1081	-5.8371	-5.1428	-5.8627	-4.0828	α/SF
	12.256[b]	-3.2481	-2.8174	-2.3297	-3.0234		
$^{318}122$	13.114[a]	-5.6685	-5.3884	-4.6988	-5.4239	-13.2452	SF
	10.642[b]	0.4301	0.8659	1.1253	0.4588		
$^{328}126$	14.424[a]	-7.2231	-6.9811	-6.2090	-6.9609	-0.8647	α
	12.641[b]	-3.5790	-3.1616	-2.6497	-3.3709		
$^{324}124$	14.070[a]	-7.0307	-6.8318	-6.0410	-6.8043	-11.6585	SF
	11.997[b]	-2.6240	-2.2072	-1.7331	-2.4562		

[a] Q-value calculated using binding energy obtained from DIRHBS code(with DD-PC1)

[b] Q-value calculated using binding energy obtained from the WS4 mass table.

Table 3.16: The Q_α, $T_{1/2}$ and decay mode of isotopes derived in the α-decay chain of $^{350}138$ and $^{352}138$

Parent Nuclei	Q (MeV)	$log_{10}T_{1/2}$ (alpha decay)				$log_{10}T_{1/2}$ (fission)	Decay Mode
		Model	UDL	Royer	VS		
$^{350}138$	15.9633 [a]	-6.9909	-6.6962	-6.1908	-6.7818	110.3197	α
	15.4215 [b]	-5.9178	-5.6793	-5.2413	-5.8277		
$^{346}136$	15.2575 [a]	-6.1342	-5.8887	-5.3953	-6.0084	87.9126	α
	15.1645 [b]	-5.9496	-5.7084	-5.2271	-5.8392		
$^{342}134$	15.0482 [a]	-5.76183	-6.0131	-5.4720	-6.1093	67.6918	α
	14.7738 [b]	-5.0788	-5.4735	-4.9684	-5.6028		
$^{338}132$	15.0045 [a]	-6.7823	-6.4598	-5.8512	-6.5128	49.5569	α
	14.3920 [b]	-5.4388	-5.2474	-4.7203	-5.3744		
$^{334}130$	15.0019 [a]	-7.1750	-6.9843	-6.3049	-6.9896	33.4089	α
	13.9555 [b]	-5.0077	-4.8976	-4.3591	-5.0295		
$^{330}128$	14.9635 [a]	-7.5322	-7.4406	-6.6969	-7.4025	19.1502	α
	12.8957 [b]	-3.5035	-3.1280	-2.6766	-3.3501		
$^{326}126$	14.7167 [a]	-7.7204	-7.5084	-6.7287	-7.4502	6.6844	α
	12.7736 [b]	-3.6283	-3.4413	-2.9387	-3.6274		
$^{322}124$	13.5958 [a]	-6.1081	-5.8371	-5.1428	-5.8627	-4.0828	α/SF
	12.5236 [b]	-3.8660	-3.4580	-2.9266	-3.6258		
$^{318}122$	13.1144 [a]	-5.6685	-5.3884	-4.6988	-5.4239	-13.2452	SF
	11.5131 [b]	-1.9676	-1.5656	-1.1390	-1.8282		
$^{352}138$	15.5756 [a]	-6.1829	-6.0040	-5.5164	-6.1385	102.9266	α
	15.8961 [b]	-6.7076	-6.6032	-6.0758	-6.7007		
$^{348}136$	14.8112 [a]	-5.4635	-5.0387	-4.5738	-5.2169	80.4937	α
	14.5966 [b]	-4.9601	-4.6012	-4.1654	-4.8063		
$^{344}134$	14.5570 [a]	-5.4024	-5.0666	-4.5606	-5.2271	60.2471	α
	14.3175 [b]	-4.9848	-4.5723	-4.0993	-4.7630		
$^{340}132$	14.4984 [a]	-5.8514	-5.4939	-4.9219	-5.6121	42.0863	α
	14.0282 [b]	-4.7080	-4.5204	-4.0139	-4.6980		
$^{336}130$	14.5031 [a]	-6.3550	-6.0482	-5.4037	-6.1165	25.9122	α
	13.5338 [b]	-4.4460	-4.0195	-3.5119	-4.2110		
$^{332}128$	14.5021 [a]	-6.2583	-6.5892	-5.8749	-6.6087	11.6272	α
	13.0651 [b]	-3.9727	-3.5498	-3.0415	-3.7528		
$^{328}126$	14.4249 [a]	-7.2231	-6.9811	-6.2090	-6.9609	-0.8647	α
	12.3808 [b]	-2.9979	-2.5350	-2.0659	-2.7820		
$^{324}124$	14.0700 [a]	-7.0307	-6.8318	-6.0410	-6.8043	-11.6585	SF
	11.9127 [b]	-2.4031	-1.9916	-1.5323	-2.2535		

[a] Q-value calculated using binding energy obtained from DIRHBS code(with DD-PC1)
[b] Q-value calculated using binding energy obtained from the WS3.3 mass table.

Table 3.17: Same as in Table 3.16 for $^{354-358}138$

Parent Nuclei	Q (MeV)	$log_{10}T_{1/2}$ (alpha decay)				$log_{10}T_{1/2}$ (fission)	Decay Mode
		Model	UDL	Royer	VS		
$^{354}138$	15.1979 [a]	-5.7437	-5.3039	-4.8347	-5.4878	94.8215	α
	15.4086 [b]	-6.1003	-5.7146	-5.2181	-5.8730		
$^{350}136$	14.3795 [a]	-4.2319	-4.1788	-3.7431	-4.4161	72.3628	α
	14.8634 [b]	-5.3275	-5.1739	-4.6719	-5.3499		
$^{346}134$	14.0722 [a]	-4.5429	-4.0829	-3.6146	-4.3098	52.0904	α
	13.7477 [b]	-3.8349	-3.3746	-2.9538	-3.6450		
$^{342}132$	13.9901 [a]	-4.7387	-4.4695	-3.9383	-4.6564	33.9036	α
	13.6154 [b]	-4.1100	-3.6546	-3.1782	-3.8912		
$^{338}130$	13.9901 [a]	-5.4107	-5.0310	-4.4270	-5.1673	17.7035	α
	13.3156 [b]	-3.6916	-3.5624	-3.0576	-3.7880		
$^{334}128$	14.0028 [a]	-5.7556	-5.6166	-4.9401	-5.7011	3.3924	α
	12.5360 [b]	-2.8055	-2.3313	-1.8776	-2.6143		
$^{330}126$	13.9713 [a]	-6.3626	-6.1096	-5.3687	-6.1481	-9.1257	SF
	12.2638 [b]	-2.5595	-2.2777	-1.7980	-2.5465		
$^{356}138$	14.8336 [a]	-4.9996	-4.6033	-4.1527	-4.8366	86.0044	α
	15.2294 [b]	-5.7629	-5.3957	-4.8925	-5.5800		
$^{352}136$	13.9667 [a]	-3.8237	-3.3191	-2.9129	-3.6155	63.5201	α
	14.3222 [b]	-4.0012	-4.0874	-3.6299	-4.3365		
$^{348}134$	13.5988 [a]	-3.3083	-3.0710	-2.6426	-3.3662	43.2219	α
	13.7206 [b]	-3.8040	-3.3442	-2.8975	-3.6227		
$^{344}132$	13.4855 [a]	-3.2750	-3.3941	-2.9073	-3.6529	25.0093	α
	13.2101 [b]	-2.6275	-2.7644	-2.3199	-3.0616		
$^{340}130$	13.4726 [a]	-4.3850	-3.9440	-3.3855	-4.1526	8.7832	α
	12.9096 [b]	-3.1557	-2.6535	-2.1822	-2.9406		
$^{336}128$	13.4874 [a]	-4.8560	-4.5536	-3.9211	-4.7086	-5.5539	SF
	12.2992 [b]	-2.0786	-1.7764	-1.3325	-2.0993		
$^{358}138$	14.4865 [a]	-3.9622	-3.9113	-3.4789	-4.1935	76.4757	α
	14.8104 [b]	-4.9934	-4.5856	-4.1084	-4.8261		
$^{354}136$	13.5756 [a]	-2.9873	-2.4689	-2.0917	-2.8239	53.9658	α
	14.3825 [b]	-4.6576	-4.2446	-3.7489	-4.4901		
$^{350}134$	13.1417 [a]	-2.5548	-2.0420	-1.6547	-2.4065	33.6418	α
	13.3200 [b]	-2.2860	-2.4613	-2.0460	-2.8001		
$^{346}132$	12.9889 [a]	-2.7548	-2.2739	-1.8347	-2.6073	15.4033	α
	13.0448 [b]	-2.9381	-2.4066	-1.9585	-2.7319		
$^{342}130$	12.9562 [a]	-3.1876	-2.7934	-2.2848	-3.0782	-0.8486	α
	12.4306 [b]	-1.8016	-1.5170	-1.0947	-1.8794		
$^{338}128$	12.9652 [a]	-3.8221	-3.4100	-2.8273	-3.6404	-15.2119	SF
	11.9186 [b]	-1.3629	-0.8299	-0.4225	-1.2164		

[a] Q-value calculated using binding energy obtained from DIRHBS code(with DD-PC1)
[b] Q-value calculated using binding energy obtained from the WS3.3 mass table.

Table 3.18: Same as in Table 3.16 for $^{360-370}138$

Parent Nuclei	Q (MeV)	$log_{10}T_{1/2}$ (alpha decay)				$log_{10}T_{1/2}$ (fission)	Decay Mode
		Model	UDL	Royer	VS		
$^{360}138$	14.1608 [a]	-3.0867	-3.2398	-2.8244	-3.5696	66.2355	α
	14.6735 [b]	-4.6262	-4.3329	-3.8448	-4.5950		
$^{356}136$	13.2089 [a]	-1.9739	-1.6382	-1.2888	-2.0505	43.7000	α
	13.9982 [b]	-3.4155	-3.4477	-2.9775	-3.7485		
$^{352}134$	12.7052 [a]	-1.3167	-1.0074	-0.6619	-1.4416	23.3503	α
	13.4650 [b]	-3.3548	-2.8260	-2.3586	-3.1486		
$^{348}132$	12.5050 [a]	-1.6405	-1.1176	-0.7286	-1.5278	5.08607	α
	12.6237 [b]	-1.6865	-1.4147	-1.0057	-1.8068		
$^{344}130$	12.4455 [a]	-0.8268	-1.5839	-1.1295	-1.9485	-11.1918	SF
	12.1234 [b]	-1.0710	-0.7626	-0.3638	-1.1772		
$^{362}138$	13.8612 [a]	-2.2851	-2.6027	-2.2022	-2.9780	55.2839	α
	14.2704 [b]	-4.0349	-3.5079	-3.0472	-3.8272		
$^{358}136$	12.8695 [a]	-0.7408	-0.8390	-0.5155	-1.3067	32.7228	α
	13.9953 [b]	-3.3071	-3.4708	-2.9715	-3.7761		
$^{354}134$	12.2948 [a]	-0.6187	0.0149	0.3193	-0.4882	12.3475	α
	13.1393 [b]	-2.6380	-2.0952	-1.6493	-2.4688		
$^{350}132$	12.0385 [a]	-0.3486	0.0635	0.4004	-0.4251	-5.9424	SF
	12.7624 [b]	-2.2922	-1.7861	-1.3246	-2.1617		
$^{364}138$	13.6017 [a]	-2.4191	-2.0368	-1.6466	-2.4532	43.6212	α
	13.7956 [b]	-2.8878	-2.4829	-2.0630	-2.8717		
$^{360}136$	12.5619 [a]	-0.2277	-0.0882	0.2124	-0.6084	21.0346	α
	13.6801 [b]	-2.5548	-2.7961	-2.3145	-3.1491		
$^{356}134$	11.9147 [a]	0.3929	1.0079	1.2730	0.4376	0.6337	α
	13.6801 [b]	-2.5548	-2.7961	-2.3145	-3.1491		
$^{352}132$	11.5944 [a]	0.6839	1.2538	1.5378	0.6861	-17.6819	SF
	13.6801 [b]	-2.5548	-2.7961	-2.3145	-3.1491		
$^{366}138$	13.4325 [a]	-2.2702	-1.6686	-1.2757	-2.1139	31.2475	α
	14.1862 [b]	-3.8872	-3.3833	-2.8762	-3.7222		
$^{362}136$	12.2982 [a]	0.0093	0.5744	0.8579	0.0072	8.6354	α
	13.0905 [b]	-1.4248	-1.4406	-1.0224	-1.8834		
$^{358}134$	11.5714 [a]	1.3382	1.9450	2.1743	1.3109	-11.7909	SF
	13.0158 [b]	-2.3075	-1.8579	-1.3734	-2.2582		
$^{368}138$	14.7270 [a]	-5.0390	-4.5607	-3.9482	-4.8326	18.1630	α
	14.0109 [b]	-3.4721	-3.0259	-2.5155	-3.3930		
$^{364}136$	12.1300 [a]	0.3984	0.9985	1.2807	0.3991	-4.4744	SF
	13.4369 [b]	-2.7883	-2.2943	-1.7920	-2.6904		
$^{370}138$	14.6305 [a]	-4.7181	-4.3896	-3.7614	-4.6779	4.3678	α
	13.6986 [b]	-2.8601	-2.3481	-1.8559	-2.7632		
$^{366}136$	13.3908 [a]	-2.6383	-2.2153	-1.6912	-2.6222	-18.2950	SF
	13.6884 [b]	-3.3703	-2.9022	-2.3321	-3.2666		

[a] Q-value calculated using binding energy obtained from DIRHBS code(with DD-PC1)
[b] Q-value calculated using binding energy obtained from the WS3.3 mass table.

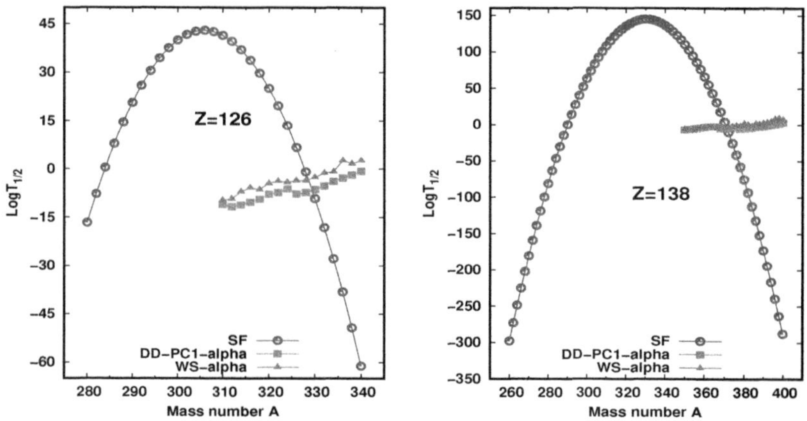

Figure 3.22: Comparison of alpha decay and SF half-live values for Z=126 and Z=138 isotopes is shown.

DD-PC1 binding energy calculations as in fig. 3.18. In fig. 3.22, the alpha decay and SF half-life are shown as a function of mass number which brings out the mass range in which the alpha decay chain is possible.

3.4 Decay properties of hyperheavy nuclei $Z = 164$

In some astrophysical processes like supernova explosion and neutron star merger where neutron flux is in abundance, SHN and hyperheavy nuclei may get formed, but there is no direct evidence of SHN as of now. The spectra of the Przybylski's star has given evidence for many actinide atoms and ions up to Einsteinium(Z = 99) (Buck et al., 1991). Also, in the OLIMPYIA project the charge distribution studies in olive crystal of meteorites had given evidence for three SHN in the range $105 < Z < 130$ and further regression analysis had provided an accurate estimate of the Z of one of these nuclei to be 119 ± 6 with a probability of 95 percent. All these findings are through theoretical studies of properties of SHN and their stability, hence theoretical half-life studies have considerable importance. In this part, a systematic study on decay properties of Z = 164 nucleus, which is expected to be the next proton magic number, around which a magic island of hyper-heavy element could occur is presented. For the present study, the DIRHB computer code have been used which is a relativistic self-consistent mean-field framework code

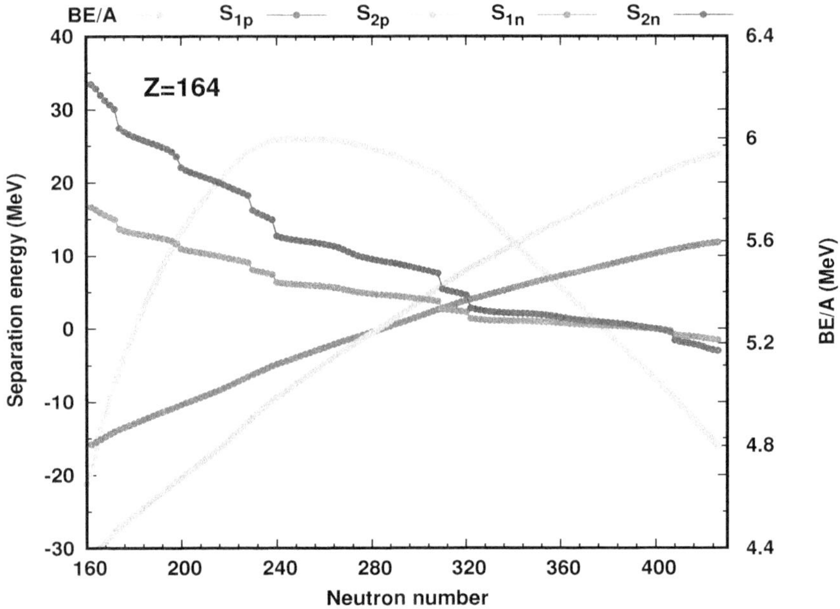

Figure 3.23: The 1p, 1n, 2p & 2n and BE/A of Z=164 isotope

which solves the stationary relativistic Hartree-Bogoliubov equations. Using DD-PC1 effective interaction the ground state properties of nuclei are obtained.

The BE/A curve in fig. 3.23 reveals that isotopes with N = 240-260 are tightly bound. This 1p, 1n, 2p and 2n separation energy curves bring the isotopes which are prone to proton and neutron decay (the ones below $S_{p(n)} = 0$). The α-decay half-life presented in fig. 3.24 consists of several peaks which suggest relative stability of the nuclide compared to neighboring isotopes, which indicates the possible neutron magic/sub-magic numbers at N = 172, 184, 190, 198, 228, 238, 320, 404 & 420. In fig. 3.24 the 1p, α and SF half-lives are also shown. The life time with smaller $T_{1/2}$ dominates the other modes of decay. The proton decay is the dominant decay mode in the range $326 \leq A \leq 392$, and α-emission in $394 \leq A \leq 470$, and SF in $472 \leq A \leq 590$. The dashed lines indicate the experimental detection limits $10^{-6} \leq T_{1/2} \leq 10^{30}$ s, and isotopes lying in this region of experimental limits is only considered in this work. The crossing over of 1p and α decay suggests that the next neutron magic number beyond 184 will be around 228. These results may be highly useful in studying SHN traces in astrophysical and cosmic sources, and for

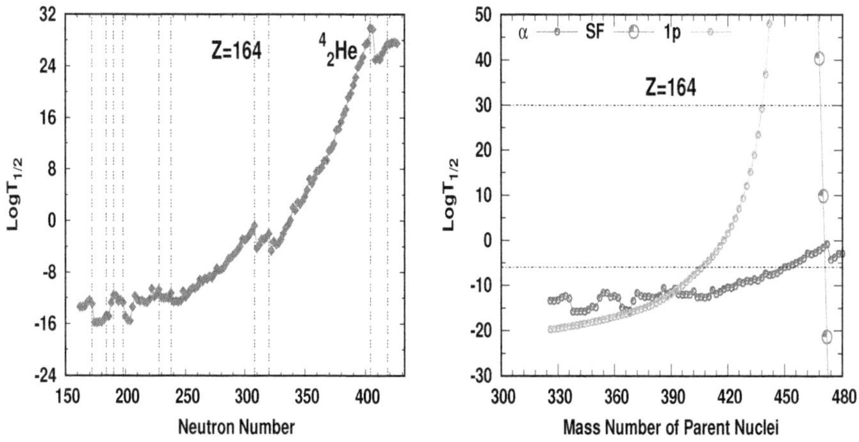

Figure 3.24: The α-decay half life of Z=164 isotopes.

upcoming synthesis through experiments.

3.5 Summary

To enumerate, the decay modes of even-even isotopes for $Z = 130 - 144$, the alpha decay chains at $Z = 122$, $Z = 126$ and $Z = 138$ using a fission like approach evaluated with Improved Transfer-matrix method employing Cubic Plus Proximity Potential provide the following outcomes:

- Decay property studies on the superheavy range with $Z = 130 - 144$ in the mass range $290 \leq A \leq 380$ suggests the possible existence of neutron magic numbers at $N = 172$, 184, 198, 228, 238 and proton magic number at $Z = 138$.

- The proton shell closure at $Z = 138$ seen as a pronounced gap between the neighbouring nuclei, has a considerable effect on proton emission, alpha decay and cluster decay half-lives in the SHN.

- The neutron shell closure at $N = 172, 184, 198, 228, 238$ is seen to affect only the alpha and cluster decay half-lives, from the dips seen in the figures plotted.

- The study in this range of charge and mass number ought to have nuclei in the dominant decay modes depends upon its neutron number within the experimental constraints of $10^{-6}s \leq T_{1/2} \leq 10^{30}$ s i.e., for a given Z with increasing A the dominant decay mode is seen to shift from proton decay to α-decay and finally SF.

- The signature of such decays may have an impact in the experiments for identifying the heavier isotopes to be synthesized in laboratories in future.

- The competition between the decay modes suggests the possible decay modes to be considered while synthesizing these isotopes.

- The α-decay chains for yet-to-be-discovered SHE $Z = 122$ are studied in the mass range $298 \leq A \leq 350$, $Z = 126$ in the mass range $314 \leq A \leq 340$ and $Z = 138$ in the mass range $350 \leq A \leq 400$.

- The model incorporates important features such as improving the interaction potential and half-life predictions. The form of cubic potential used by following the conservation of energy, the effective mass in the pre-scission region and reduced mass in the post-scission region refines the dynamics of the decay process. The accuracy of the decay rates calculated is considerably increased while using improved transfer matrix method.

- To avoid ambiguity, binding energy data WS4, DD-PC1, WS3.3 are employed in this study. The Q-values extracted from WS-4 mass model for $Z = 122$, WS4, DD-PC1, WS3.3 for proton shell closure $Z = 126$ & $Z = 138$ suggests least error in binding energy of SHN than other mass models.

- The calculated half-life using CPP-ITM model is in good agreement with empirical formula values. The decay chains from isotopes $^{298-300,302,304-306,308-310,312,314}122$ is likely to contain 7α, 5α, 4α, 3α, 2α and 1α decays. For $^{314,316\&318,320,322\&324,326,328}126$ emit 6,5,4,3,2 and 1α and $^{350,352,354,356\&358,360,362\&364,366,368\&370}138$ emit 8,7,6,5,4,3,2 and 1α emission.

- It is to be highlighted here that, the isotopes of $Z = 122$ with $A > 350$, $Z = 126$ with $A > 328$ and $Z = 138$ with $A > 370$ decay via SF.

- From the study on $Z = 164$ it is predicted that $N = 228$ may be the next magicity.

The theoretical predictions reported here will certainly have useful guidances for the future experimental efforts towards the synthesis and(or) identification of these superheavy isotopes.

References

Agbemava, S. E., Afanasjev, A. V., Ray, D. and Ring, P. (2014). Global performance of covariant energy density functionals: Ground state observables of even-even nuclei and the estimate of theoretical uncertainties. *Physical Review C*, **89**, 054320.

Bhuyan, M. and Patra, S. (2012). Magic nuclei in superheavy valley. *Modern Physics Letters A*, **27**, 1250173.

Biswas, D. and Kumar, V. (2014). Improved transfer matrix methods for calculating quantum transmission coefficient. *Physics Review E*, **90**, 013301.

Buck, B., Merchant, A. C. and Perez, S. M. (1991). Recent developments in the theory of alpha decay. *Modern Physics Letters A*, **06**, 2453-2461.

Carmel Vigila Bai, G. M. and Agnes, R. N. (2017). Alpha decay and cluster decay of some neutron-rich actinide nuclei. *Pramana – Journal of Physics*, **88**, 43.

Cui, J. P., Zhang, Y. L., Zhang, S. and Wang, Y. Z. (2018). α-decay half-lives of superheavy nuclei. *Physical Review C*, **97**, 014316.

Dong, T. and Ren, Z. (2005). New model of binding energies of heavy nuclei with $Z \geq 90$. *Physical Review C*, **72**, 064331.

Dong, T. and Ren, Z. (2008). Improved version of a binding energy formula for heavy and superheavy nuclei with $Z \geqslant 90$ and $N \geqslant 140$. *Physical Review C*, **77**, 064310.

Gambhir, Y., Bhagwat, A. and Gupta, M. (2009). The superheavy elements: overview. *Indian Journal of Physics*, **83**, 661–669.

Geiger, H. and Nuttall, J. (1911). Lvii. the ranges of the α-particles from various radioactive substances and a relation between range and period of transformation. *The London, Edinburgh, and Dublin Philosophical Magazine and Journal of Science*, **22**, 613-621.

Gherghescu, R., Poenaru, D. and Greiner, W. (1998). Energy valleys of double-magic superheavy nuclei. *Il Nuovo Cimento A (1971-1996)*, **111**, 105–111.

Gherghescu, R. A. and Greiner, W. (2012). Fusion valleys for synthesis of superheavy nuclei. *Nuclei Far from Stability and Astrophysics*, **17**, 197.

Guo, C. L. and Zhang, G. L. (2014). Analysis of proton radioactivity of nuclei by using proximity potential with a new universal function. *The European Physical Journal A*, **50**, 187.

Hofmann, S., Münzenberg, G. and Schadel, M. (2004). Feature article: On the discovery of superheavy elements. *Nuclear Physics News*, **14**, 5-13.

Jackson, K., Cardinal, C., Evans, H., Jelley, N. and Cerny, J. (1970). 53com: A proton-unstable isomer. *Physics Letters B*, **33**, 281.

Krane, K. S. (1988). *Introductory nuclear physics*. New York, NY: Wiley.

Lalazissis, G. A., Nikšić, T., Vretenar, D. and Ring, P. (2005). New relativistic mean-field interaction with density-dependent meson-nucleon couplings. *Physical Review C*, **71**, 024312.

Manjunatha, H. C. and Sowmya, N. (2018). Decay modes of superheavy nuclei Z = 124. *International Journal of Modern Physics E*, **27**, 1850041.

Manjunatha, H. C., Sridhar, K. N. and Sowmya, N. (2018). Investigations of the synthesis of the superheavy element Z=122. *Physical Review C*, **98**, 024308.

Moller, P., Sierk, A., Ichikawa, T. and Sagawa, H. (2016). Nuclear ground-state masses and deformations: Frdm(2012). *Atomic Data and Nuclear Data Tables*, **109-110**, 1.

Ni, D. and Ren, Z. (2012). New formula of half-lives for proton emission from spherical and deformed nuclei. *Romanian Journal of Physics*, **57**, 407-418.

Nikšić, T., Paar, N., Vretenar, D. and Ring, P. (2014). Dirhb—a relativistic self-consistent mean-field framework for atomic nuclei. *Computer Physics Communications*, **185**, 1808.

Nikšić, T., Vretenar, D. and Ring, P. (2008). Relativistic nuclear energy density functionals: Adjusting parameters to binding energies. *Physical Review C*, **78**, 034318.

Oganessian, Y., Abdullin, F., Dmitriev, S., Itkis, M., Polyakov, A., Sagaidak, R., ... Rykaczewski, K. (2013). New insights into the discoveries of elements 113, 115 and 117. In *Exotic nuclei* (p. 163-166).

Oganessian, Y. T. (2007). Heaviest nuclei from ^{48}Ca-induced reactions. *Journal of Physics G: Nuclear and Particle Physics*, **34**, R165–R242.

Oganessian, Y. T., Utyonkov, V. K., Lobanov, Y. V., Abdullin, F. S., Polyakov, A. N., Sagaidak, R. N., ... Lougheed, R. W. (2006). Synthesis of the isotopes of elements 118 and 116 in the ^{249}Cf and ^{245}Cm $+^{48}$ Ca fusion reactions. *Physical Review C*, **74**, 044602.

Patra, S. K., Wu, C.-L., Praharaj, C. R. and Gupta, R. K. (1999). A systematic study of superheavy nuclei for Z = 114 and beyond using the relativistic mean field approach. *Nuclear Physics*, **651**, 117-139.

Poenaru, D. N. and Gherghescu, R. A. (2014). Fission decay of ^{282}Cn studied using cranking inertia. *Journal of Physics G: Nuclear and Particle Physics*, **41**, 125104.

Poenaru, D. N. and Gherghescu, R. A. (2016). Spontaneous fission of the superheavy nucleus ^{286}Fl. *Physical Review C*, **94**, 014309.

Poenaru, D. N. and Gherghescu, R. A. (2017). Cluster preformation at the nuclear surface in cold fission. *Europhysics Letters*, **118**, 22001.

Poenaru, D. N. and Gherghescu, R. A. (2018). α decay and cluster radioactivity of nuclei of interest to the synthesis of Z=119, 120 isotopes. *Physical Review C*, **97**, 044621.

Qi, C., Xu, F. R., Liotta, R. J. and Wyss, R. (2009). Universal decay law in charged-particle emission and exotic cluster radioactivity. *Physical Review Letters.*, **103**, 072501.

Rose, H. J. and Jones, G. A. (1984). A new kind of natural radioactivity. *Nature*, **307**, 245-247.

Royer, G. (2000). Alpha emission and spontaneous fission through quasi-molecular shapes. *Journal of Physics G: Nuclear and Particle Physics*, **26**, 1149–1170.

Samanta, C. and Adhikari, S. (2002). Extension of the bethe-weizsäcker mass formula to light nuclei and some new shell closures. *Physical Review C*, **65**, 037301.

Sandulescu, A., Poenaru, D. and Greiner, W. (1980). New type of decay of heavy nuclei intermediate between fission and. cap alpha. decay. *Soviet Journal of Nuclear Physics*, **6**.

Santhosh, K. P. and Nithya, C. (2017). Decay properties of $^{256-339}$Ds superheavy nuclei. *The European Physical Journal A*, **53**, 189.

Santhosh, K. P. and Nithya, C. (2018a). Predictions on the modes of decay of even Z superheavy isotopes within the range $104 \leq Z \leq 136$. *Atomic Data and Nuclear Data Tables*, **119**, 33.

Santhosh, K. P. and Nithya, C. (2018b). Predictions on the modes of decay of odd Z superheavy isotopes within the range $105 \leq Z \leq 135$. *Atomic Data and Nuclear Data Tables*, **121-122**, 216.

Santhosh, K. P. and Priyanka, B. (2014). Predictions for the α-decay chains of Z=120 superheavy nuclei in the range $272 \leq A \leq 319$. *Physical Review C*, **90**, 054614.

Santhosh, K. P. and Priyanka, B. (2017). Predictions on the feasible alpha and cluster decays from $^{298-336}$126 superheavy nuclei. In *Nuclear particle correlations and cluster physics* (p. 383-400).

Santhosh, K. P. and Sahadevan, S. (2010). Systematic decay studies of even–even $^{132-138}$Nd, $^{144-158}$Gd, $^{176-196}$Hg and $^{192-198}$Pb isotopes. *Nuclear Physics A*, **847**, 42 - 65.

Santhosh, K. P. and Sukumaran, I. (2017a). Alpha decay studies on po isotopes using different versions of nuclear potentials. *The European Physical Journal A*, **53**, 246.

Santhosh, K. P. and Sukumaran, I. (2017b). Studies on cluster decay from trans-lead nuclei using different versions of nuclear potentials. *The European Physical Journal A*, **53**, 136.

Santhosh, K. P. and Sukumaran, I. (2018). Predictions of proton emissions from the isotopes of nuclei with Z = 100-136 using the coulomb and proximity potential model for deformed nuclei. *The European Physical Journal A*, **54**, 102.

Sharma, M. K. and Kaur, G. (2015). Cluster decay analysis and related structure effects of fissionable heavy and superheavy nuclei. *Pramana – Journal of Physics*, **85**, 431.

Utyonkov, V., Oganessian, Y. T., Dmitriev, S., Itkis, M., Moody, K., Stoyer, M., ... Hamilton, J. (2017). Discovery of elements 113-118. *Eurasian Journal of Physics and Functional Materials*, **1**, 5–11.

Viola, V. and Seaborg, G. (1966). Nuclear systematics of the heavy elements—ii lifetimes for alpha, beta and spontaneous fission decay. *Journal of Inorganic Nuclear Chemistry*, **28**, 741 - 761.

Wang, N., Liang, Z., Liu, M. and Wu, X. (2010). Mirror nuclei constraint in nuclear mass formula. *Physical Review C*, **82**, 044304.

Wang, N., Liu, M., Wu, X. and Meng, J. (2014). Surface diffuseness correction in global mass formula. *Physics Letters B*, **734**, 215.

Xu, C. and Ren, Z. (2005). Systematical law of spontaneous fission half-lives of heavy nuclei. *Physical Review C*, **71**, 014309.

Xu, C., Ren, Z. and Guo, Y. (2008). Competition between α decay and spontaneous fission for heavy and superheavy nuclei. *Physical Review C*, **78**, 044329.

Zhang, W., Meng, J., Zhang, S., Geng, L. and Toki, H. (2005). Magic numbers for superheavy nuclei in relativistic continuum hartree–bogoliubov theory. *Nuclear Physics A*, **753**, 106–135.

Zhang, Y. and Wang, Y. (2017). Systematic study on the competition between α-decay and spontaneous fission of superheavy nuclei. *Nuclear Physics A*, **966**, 102-112.

CHAPTER 4

TEMPERATURE AND DEFORMATION DETERMINED DECAY PROPERTIES

4.1 Introduction

The interaction between the nucleons governs the various nuclear properties ranging from nuclear radius, density, magnetic moment, etc (Kaur et al., 2018). The shape of nuclei is one of its fundamental property. The microscopic viewpoint considers nuclei as a quantum system which is an ensemble of protons and neutrons bound by nuclear force, whereas in the macroscopic viewpoint it is similar to that of a liquid-drop. Atoms attain their shape due to the potential experienced by electrons because of the electromagnetic force of the nucleus. However, the nuclear shape is due to the interaction between the nucleons. The nucleus can attain different shapes such as spherical, axial symmetric shapes like the prolate and oblate, and also triaxial, etc., based on shell effects (Görgen et al., 2009). The nuclei with shell closure (proton, neutron, or both) are comparatively stable and are also seen to possess spherical shapes. The partially filled shells induce asymmetrical mass distribution and show deviations from the spherical shape. The shape of nuclei though in general varies with the number of nucleons, they are also seen to vary due to its excitation energy or angular momentum. In some cases, configurations corresponding to different shapes coexist at similar energies called the 'shape co-existence'and this was first predicted by Morinaga (Morinaga, 1956). Morinaga et al., predicted that 0^+ state in ^{16}O was deformed while the ground state had a spherical shape (Morinaga, 1956). The shape co-existence is now considered widespread, and that several spherical shaped nuclei in its ground state deviate from their original form in low lying excited states. Nuclear shape-related experimental measurements can provide the details on quadrupole moments of the states and electromagnetic transition rates between them. The spherical symmetry of the nucleus is maintained around Z=50, 82 and N=82, 126 magic numbers in heavy nuclei. It is expected that the spherical symmetry and shell closures may also occur around N=172, 184, and Z=120 or 126 locations in superheavy nuclei (Myers and Swiatecki, 1966; Ghodsi and Lari, 2014; Kruppa et al., 2000).

The fusion reactions conducted at the laboratory and those occurring in nature are influenced by the target and projectile combinations. The fusion barrier for spherical nuclei will be independent of the orientations of the nucleus involved, but in the case of deformed nuclei, the fusion barrier height is governed by both the deformation and orientation together. A deformed nuclei studied under different orientation angles can lead to different fusion barrier heights, and so orientation related studies have received considerable importance. The synthesis of superheavy nuclei is carried out with actinide targets and light projectiles that are deformed. The interaction potential between two nuclei in general depend on their shapes, the possible shape combinations of daughter-cluster nuclei are spherical-spherical, spherical-deformed, deformed-spherical, and deformed-deformed. Furthermore, each shape combination's interaction potential are unique, influencing the fusion dynamics and decay process in different manner. Also, the fusion reactions between deformed nuclei F, Ne and Mg occurring in stars, play a significant role in knowing the history of stars, nuclear burning and formation of heavy elements (Ismail and Adel, 2011; Ismail and Seif, 2011). Deformation and orientation can enhance or lower the probability of compound nucleus formation and also in the decay process, where depending on orientation angle and value of deformation parameters can further enhance or suppress the decay probability. The deformations of nuclei in light and medium mass region are small but as when the mass increases they show considerable changes in the heavy and superheavy region, therefore it is necessary for one to take into account the nuclear deformations. Several previous models involving different formalisms have only the daughter deformation or the daughter- cluster deformations or the parent and daughter deformation (Santhosh and Joseph, 2002a; Pahlavani and Rahimi Shamami, 2020; Soylu and Evlice, 2015). For the decay process in heavy and superheavy isotopes the deformations of the parent, daughter and cluster nuclei are essential.

Another important quantity that governs the surface properties and excited state of liquid drop and nucleus is its temperature. The concept of nuclear temperature came into the picture about 85 years back in the seminal works of Bethe (Bethe, 1936) and Weisskopf (Weisskopf, 1937), intending to understand the formation of the compound nucleus and its decay in reactions induced by light projectiles like neutrons, which was further extended to reactions having high energy projectiles and heavy ions. Some authors have considered the thermal effects of the compound nucleus and reported a higher estimate of decay half-life (Ghorbani et al., 2020; Zanganeh and Wang, 2014; Zanganah et al., 2020; Daei-Ataollah et al., 2018). The decay of hot and rotating compound nuclei produced in heavy-ion reactions was studied by Gupta et al., employing the dynamical cluster model (Gupta, Balasubramaniam, Kumar, Singh, Beck and Greiner, 2005). The half-lives associated with decays originating from the ground and excited states of *Ba* isotopes were studied by Santhosh et al., using coulomb plus proximity potential model (Santhosh et al., 2016). On the empirical formula part, Davidson et al., (Davidson et al., 1994) calculated the temperature-dependent separation energy of 300 nuclei and Royer and Mignen (Royer and Mignen, 1992) obtained temperature-dependent sharp radius and surface width. Introducing the concept of nuclear temperature provides a proper understanding of the mechanism of formation, decay, and instability of nuclear system (Keli et al., 2006; S. S. Hosseini et al., 2016). Alpha decay plays a crucial role in the areas of nuclear & astrophysics (in the stellar nucleosynthesis) and also remains an important decay mode for unstable nuclei (heavy nuclei to SHN) (Hofmann et al., 2007). Furthermore, investigation of the α-nuclear system has provided valuable information on the details of nuclear structure, effective nuclear interactions between daughter and α-fragment, accurate data of Q_α, decay half-life values, nuclear instability, shell effects, nuclear-spin and deformations (Oganessian et al., 1999; Oganessian, Utyonkoy et al., 2004; Oganessian, Utyonkov et al., 2004b; Oganessian, Utyonkov, Lobanov et al., 2005; Oganessian, Utyonkov, Dmitriev et al., 2005; Oganessian et al., 2006; Hofmann et al., 2007; Oganessian et al., 2007; Morita et al., 2007; Pei et al., 2007; Andreyev et al., 1999; Daei-Ataollah et al., 2018). To date, experiments have successfully synthesized transuranium elements up to 118 either by hot fusion or cold fusion reactions (Oganessian and Utyonkov, 2015; Oganessian, Utyonkov et al., 2004a), and alpha decay is observed to be the prominent mode of decay for heavy and superheavy nuclei. Major theoretical and experimental works are focused on various quantities that influence the nuclear properties such as nuclear deformation, spin, isospin and decay lifetimes. In this chapter, focus is given to two important quantities, the nuclear temperature, and deformation, and their influence on the interaction potential and decay half-lives. In the first part of this chapter focus is given to study the effects of nuclear temperature on super heavy element lifetimes with emphasis on understanding the model parameters. The second part presents the role of deformation and orientation on interaction potential and decay half-lives. The quadrupole (β_2) deformation, hexadecapole deformation (β_4), and hexacontatetrapole deformations (β_6) are considered to include the shape effects of higher-order deformation parameters. These studies could be useful in guiding the experimental works dealing with the synthesis of deformed and/or hot nuclei, and their decay properties.

4.2 Temperature dependence on decay properties

A liquid drop when attains energy and temperature can be considered to be not in its ground state. Surface tension is one such essential quantity, which represents the amount of energy or force per unit required to expand the surface. It is the tendency exhibited by fluids to attain a minimum surface area configuration, so the spherical shape. It is observed that the surface tension decreases with increasing temperature. Similar effects is seen when temperature is induced on atomic nuclei. This temperature can be thought of as that attained from the compound nucleus or the temperature associated with the decaying parent or the process itself. Hence, temperature is incorporated in the Cubic plus proximity (CPP) model (Naveya et al., 2019) as discussed in chapter 2, to determine the temperature dependence of nuclear structure and stability effects. The inclusion of temperature modifies the nuclear quantities and interaction potential and hence the decay half-lives of heavy and superheavy nuclei. The quantities which govern the nuclear potential alone depend on temperature, whereas the Coulomb potential between spherical daughter and fragment nuclei is independent of temperature. The following relations describe the temperature dependence of nuclear surface width (b) and nuclear radii (R_i) (Błocki et al., 1977; Salehi and Ghodsi, 2013),

$$b(T) = b(T=0)\left[1+0.009T^2\right], \tag{4.82}$$

$$R_i(T) = R_i(T=0)\left[1+0.0005T^2\right]. \tag{4.83}$$

From the above relations, it is evident that the radius of nuclei increases with an increase in temperature, and the change in radius given by $R_i \times 0.0005T^2$ depends on the size of nuclei, hence it is expected that heavy and superheavy nuclei will undergo a larger change in radius as compared to lighter nuclei. The temperature dependence of surface tension observed in liquids are similar as seen in nuclei. Jaqaman (Jaqaman, 1989) obtained the temperature dependence of the surface tension of a hot nucleus employing a finite-temperature generalized liquid-drop model. The relation derived was,

$$\gamma(T) = 1.4\left(1 - \frac{T}{T_c}\right)^2 MeV/fm^2. \tag{4.84}$$

To incorporate temperature in the surface energy coefficient of the proximity formalism the following relation is adopted

$$\gamma(T) = \gamma(T = 0)(1 - pT)^2, \tag{4.85}$$

where p is an adjustable parameter to provide a good description of theoretical half-life values; similar relation has been employed in ref. (Daei-Ataollah et al., 2018; Golshanian et al., 2013). For the present model involving CPP potential, $p = 0.04 - 0.2$ provides good description for temperature-dependent half-lives for $84 \leq Z \leq 118$. The temperature associated with a nucleus is related to the excitation energy of the compound nucleus E^* and is given by the relation (Le Couteur and Lang, 1959; Golshanian et al., 2013)

$$E^* = E_{kinetic} + Q = \frac{1}{a}A_pT^2 - T, \tag{4.86}$$

where $E_{kinetic} = \frac{A_d}{A_p}Q$, the value of a is taken to be 9 for intermediate-mass daughter-cluster system and is 10 for superheavy systems (Ghodsi et al., 2013). By solving the following equation numerically the temperature can be extracted

$$\frac{A_d}{A_p}Q + Q = \frac{1}{a}A_pT^2 - T. \tag{4.87}$$

The temperature dependence in the post-scission region is due to surface energy coefficient γ and nuclear radii R, whereas the temperature in the pre-scission is due to nuclear radii alone. When incorporating these changes in the interaction potential in exterior and overlapping regions it becomes temperature dependent.

$$V(r,T) = \begin{cases} V_n(r,T) + V_c(r) + V_l(r) & \text{if } r \geq r_t \\ V_{ov}(r,T) & \text{if } r < r_t \end{cases} \tag{4.88}$$

4.3 Decay properties of Z = 84-118

In the present study, decay processes of isotopes of $Z = 84 - 102$ in the heavy region (even-even) and isotopes of $Z = 104 - 118$ in the superheavy region; only those isotopes for which the experimental half-life data available are considered (S. Hosseini et al., 2018; Bao et al., 2014; Budaca and Silisteanu, 2011). Such a way, a total of 159 isotopes considered in this work are tabulated in table 4.1. To start with, a subset of heavy nuclei in the range $Z = 84 - 102$ where each charge number exists with a large number of isotopes are taken and their Q-values as a function of neutron number are shown in fig. 4.1. It is observed that the Q-value curve of Z=92 lies at the top whereas Z=84 lies at the bottom, which indicate the decay of higher atomic number nuclei results in comparatively higher energy release than the lower atomic number nuclei. The Q-value is maximum at $N = 128$ for isotopes in heavy region, which can be attributed to the existence of the neutron magic number at 126; the nuclear temperature obtained using equation (4.87) are shown in fig. 4.2 (data in table 4.1) collectively as a function of atomic number and also individually for different Z as a function of neutron number. For the studied range of nuclei, the calculated nuclear temperature varies between 0.55 and 0.95 MeV, and for A=210 to A=260 the temperature obtained is around 0.9 MeV. It can also be seen from the temperature curves fig. 4.2(b) for different Z where at N=128, the isotopes exhibit maximum temperature. From fig. 4.1 and fig. 4.2(b), a direct relationship between Q-value and temperature is seen, ie., if Q-value is large then the temperature acquired leading to decay is likely to be large.

Also the temperature has influence in the nuclear radii i.e., the radius of the nuclei shows slightly larger value for hot system than the cold system. The change in radius would have some impact on the touching distance of the daughter-cluster system, which will influence the tunneling probability and assault frequency of alpha particle. From the temperature-dependent relation of the surface energy coefficient (eqn. 4.85), it is evident that the surface energy coefficient

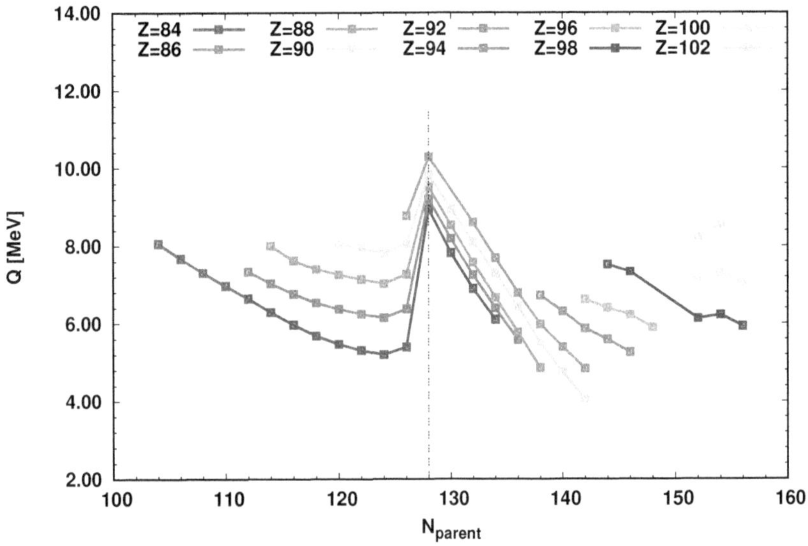

Figure 4.1: The experimental Q-values of decay processes as a function of parent nuclei neutron number.

will decrease due to the addition of temperature, which is the case for liquid drops too wherein on increasing the temperature the surface molecules get loosely bound and surface energy lowers. The temperature-independent (T-IND) and temperature-dependent (T-DEP) surface energy coefficients for Z= 84- 102 are shown in fig. 4.3. For a fixed Z, at zero temperature, γ has a linear dependence on neutron number, and it decreases on addition of neutrons. But, when the temperature is included, the linear dependence on neutron number gets altered and a big dip is seen at N=128 where $\gamma(T)$ is the lowest value since the temperature being maximum at this point. This suggests that nuclei with neutron numbers just above the closed-shell configuration (N=126) may experience a drastic reduction in surface energy and binding strength by gaining temperature. The temperature also influences the surface energy coefficient significantly compared to the nuclear radii.

The nuclear surface width for the temperature-independent case is a constant value irrespective of the size of the nuclei and other properties. The change in nuclear surface width b with temperature is represented in fig. 4.4(a). The nuclear surface width at zero temperature ie., with no temperature inclusion $b(T=0)$, is a constant value irrespective of the mass number and

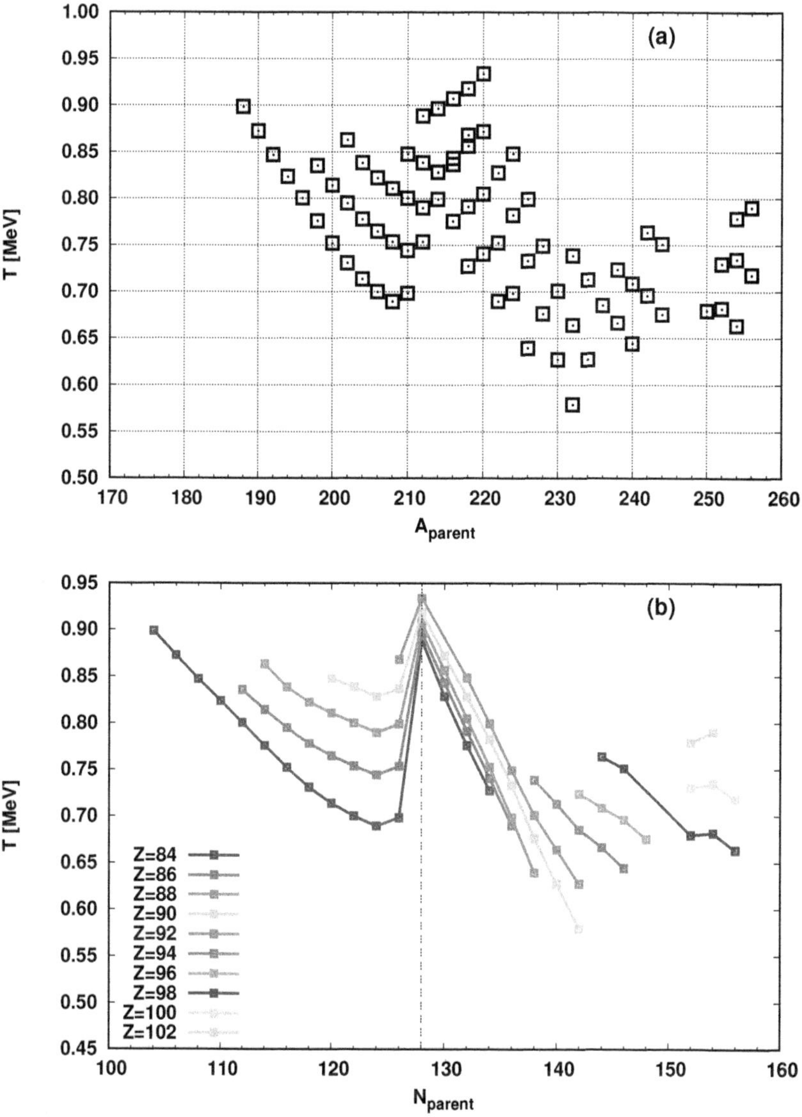

Figure 4.2: The temperature values as a function of (a) parent nuclei mass number and (b) parent nuclei neutron number in the range $84 \leq Z \leq 102$.

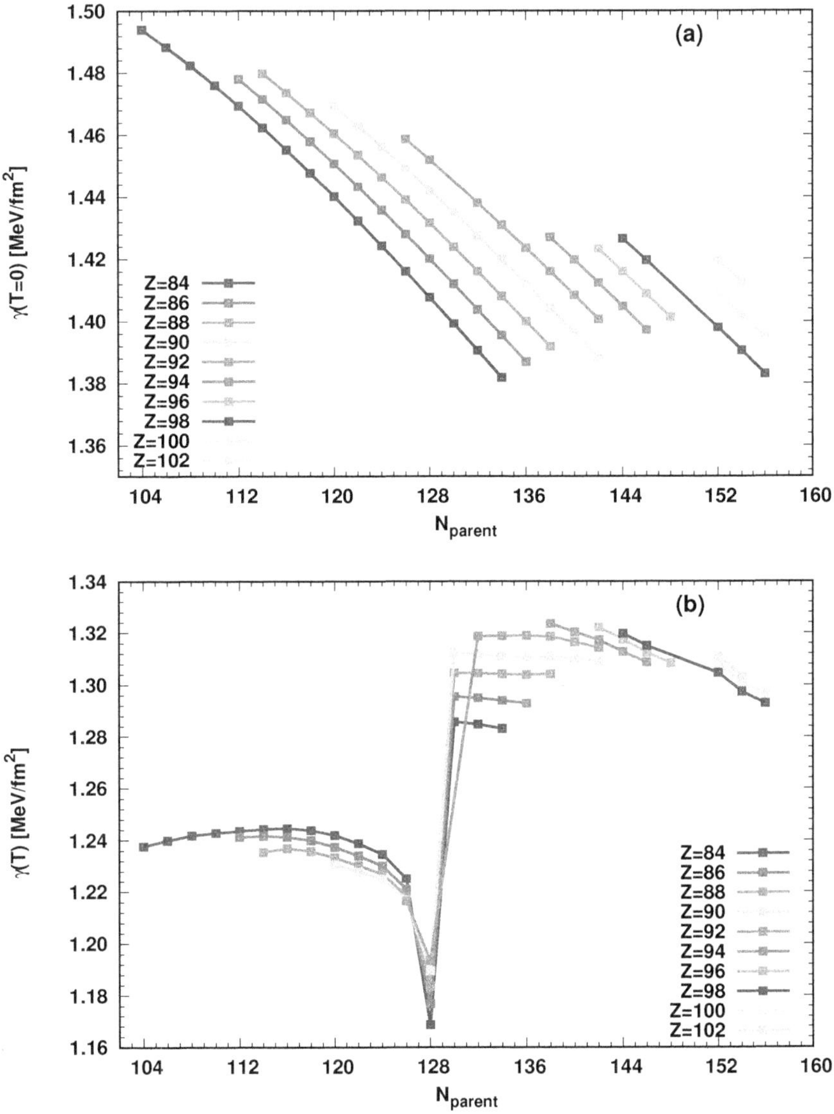

Figure 4.3: The (a) T-IND and (b) T-DEP surface energy coefficient for different isotopes of $Z = 84 - 102$.

which, attains a distribution governed by nuclear property and temperature when T $>$ 0. With the increase of temperature, the nuclear surface width is also seen to increases ($>$ 1 fm). The temperature associated with 78 superheavy isotopes Z= 105-118 are been calculated and given in table 4.1, and the results are pictorially represented in Fig. 4.4(b). The temperature values obtained for all the isotopes considered lie in between T $\approx$ 0.80 MeV and T $\approx$ 0.95 MeV, which is in correlation with the reports of Preetha and Santhosh Kumar (Preetha and Santhosh Kumar, 2020) that the collapse of pairing correlation is at T= 0.8 MeV for the hot rotating nucleus Z= 117. The temperature dependence seen in various quantities like nuclear radius, nuclear surface width, and γ will together modify the interaction potential. The temperature-independent and temperature-dependent potential barrier are drawn for the alpha decay of $^{208}Po \rightarrow ^{204}Pb + \alpha$ in fig. 4.5. It is seen from the fig. 4.5 that the barrier height and width got changed when the temperature is included ie., the width is increased due to which a small shift in radius with respect to potential is seen in the pre-scission region. The height of the barrier also seems to get increased. All these factors will have a significant effect on the penetration probability of the alpha particle. The temperature-dependent potential barrier for the decay of ^{208}Po would imply a lower tunneling probability than a temperature-independent barrier, and this will result in a higher value of half-life. In addition to the T-IND and T-DEP half-lives, the experimental half-lives and universal decay law (UDL) half-lives are also presented for comparison as seen in table 4.1.

For the isotopes of Z=84 the half-lives calculated by T-IND, T-DEP, UDL and experimental are shown in fig. 4.6(a). It is seen that the T-IND half-life curve is overlapping well with UDL values. Even, though it is close to the experimental values it does not reproduce the experimental data which lie above the T-IND curves. The inclusion of temperature through proximity formalism results a change in half-life values and the T-DEP curve overlaps well with experimental data points. The same effect is obtained for other heavy isotopes also. For Z=106 the half-lives calculated by T-IND and UDL are overlapping each other and T-DEP is very close to the experimental half-lives except at N= 163 (fig. 4.6(b)). Such good agreement of T-DEP half-life values with the experimental half-life pronounces the influence of temperature in predicting the accurate decay life time. These isotopes may have higher probability of occurrence at supernova explosion. Also $^{269}106$ is found to be an outlier since the calculated half-life

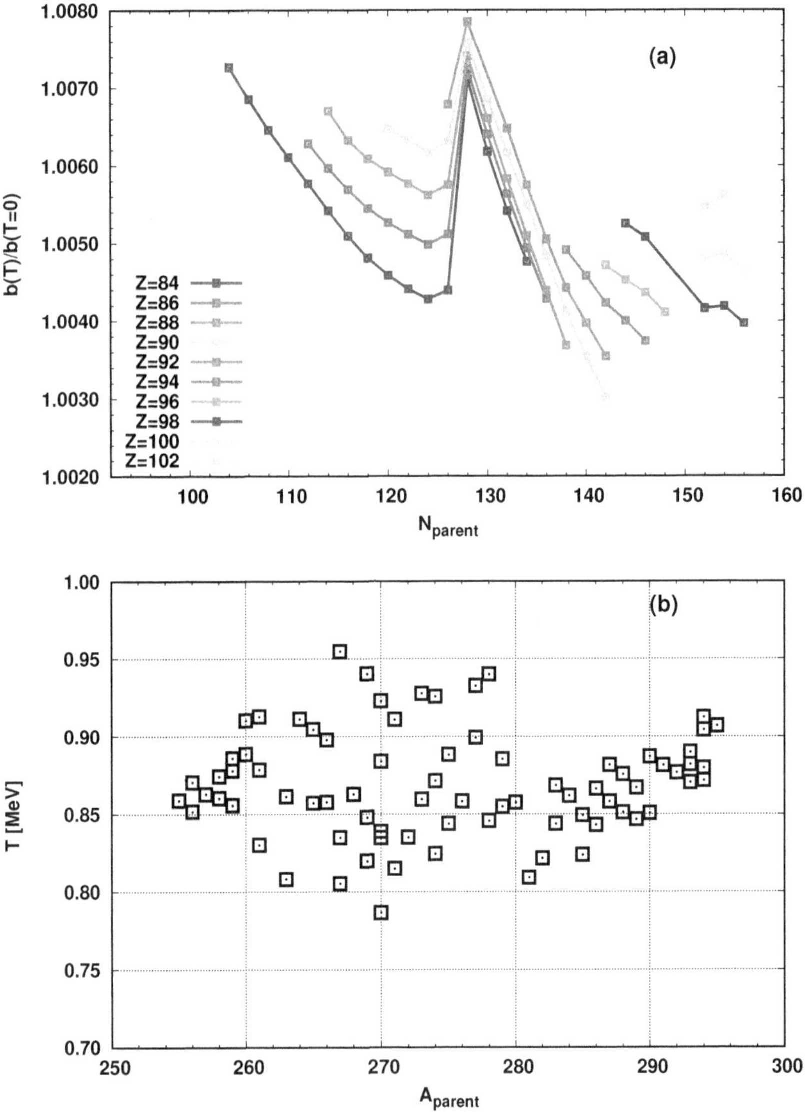

Figure 4.4: (a) The ratio of temperature dependent and temperature independent nuclear surface width for different isotopes. (b) Temperature associated with alpha decay of superheavy isotopes.

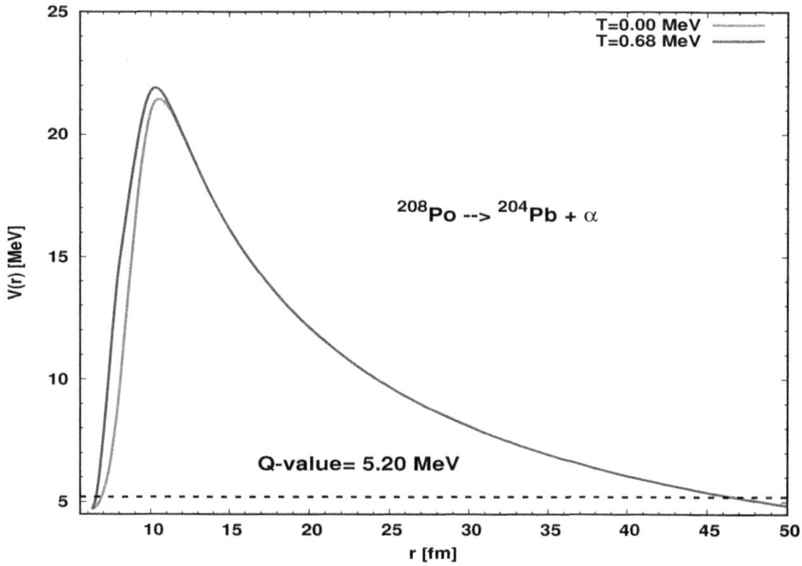

Figure 4.5: Temperature dependence of interaction potential for alpha decay of ^{208}Po; the experimental Q-value is 5.20 MeV

is comparatively less than the experimental one which suggest that this isotope might have a temperature higher than the ones calculated with the eqn. (4.87). Hence it is suggested here that the influence of temperature has a key role in the experimental decay half-lives.

In fig. 4.7, T-DEP half-life of heavy and superheavy nuclei are presented together with experimental half-life data (the black dots). It is seen that there is good agreement between the data points, (intact, overlapping), of the theoretical values and experimental data. The root mean square error (RMSE) values are found for theoretical model calculations, T-IND, T-DEP and UDL, using the formula,

$$\sigma = \sqrt{\frac{1}{n}\sum_{i=0}^{n}\left(T_{1/2}^{cal} - T_{1/2}^{exp}\right)^2}, \tag{4.89}$$

and are listed in table 4.2 for three different data sets: firstly for heavy nuclei (Z=84-102) alone, secondly for SHN (Z=104-118), and then both heavy and superheavy isotopes together. For the heavy region, it is evident that the cubic plus proximity (CPP) model has good predictive power, and in the proximity formalism, the inclusion of temperature gives better estimates of decay half-lives. Also, the T-IND model gives better estimates than Coulomb plus proximity

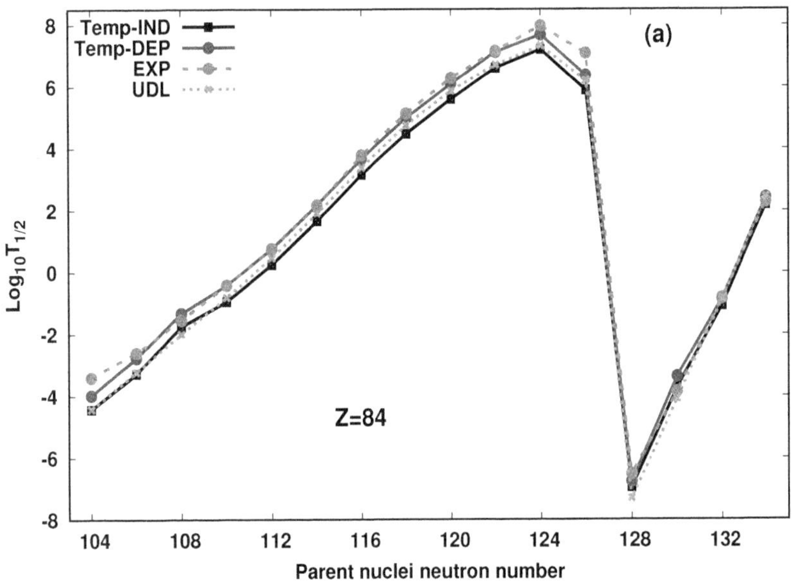

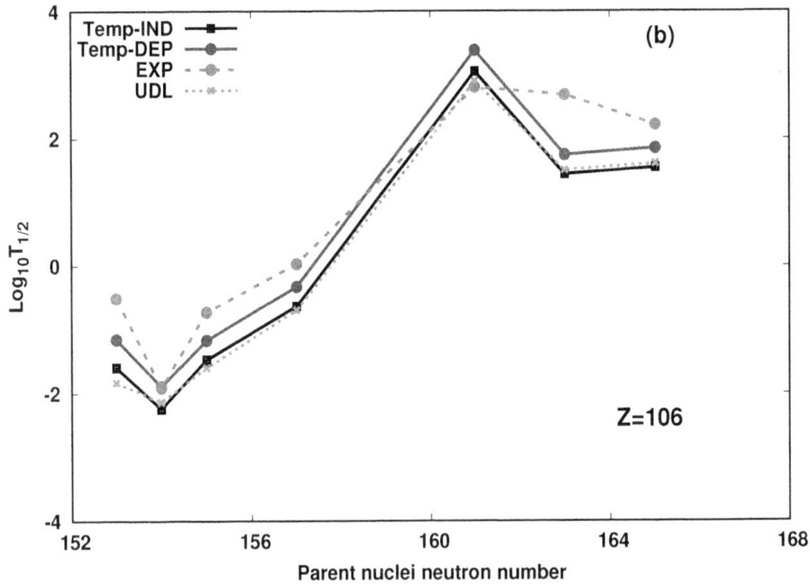

Figure 4.6: Half-lives of (a) Z=84 and (b) Z=106 isotopes calculated by T-IND, T-DEP, experimental and UDL.

model (RMSE) taken from ref. (S. Hosseini et al., 2018)). In the superheavy region, the RMSE for all formalism are higher as compared to that of the heavy region, and the T-DEP CPP model gives the lowest error among the models studied. In the third data set where both heavy and superheavy isotopes are taken, the comparatively lower RMSE for T-DEP indicates the inclusion of temperature gives a better half-life estimate. Hence it is obvious that the temperature plays an important role in radioactive decays, especially for heavy and superheavy nuclei. Also, the temperature-induced lowering of the surface energy coefficient results in comparatively increased half-life, indicating that the alpha decay could act as a probe for investigating the structure of hot compound nucleus formed in heavy-ion fusion reactions as they primarily decay by emission of alpha particles. This study may be useful for investigating the structure of hot nuclei and synthesis in laboratories.

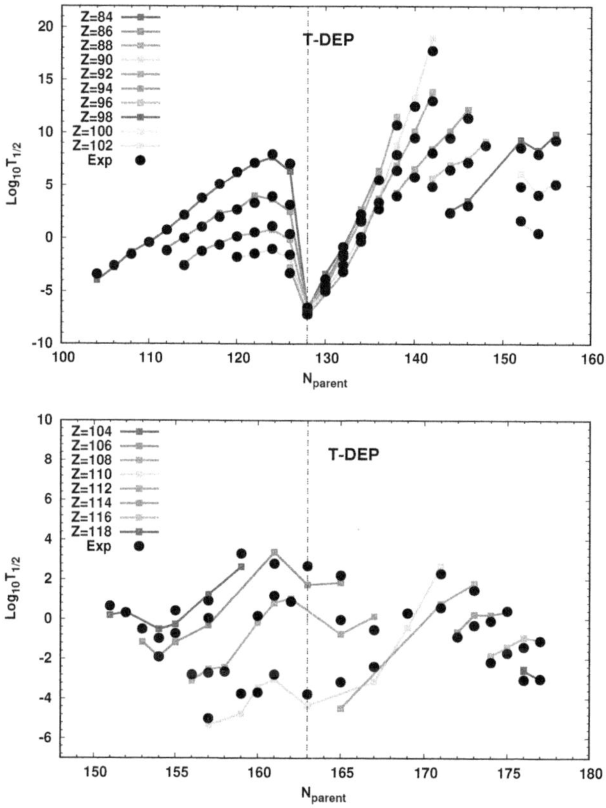

Figure 4.7: Comparison of calculated half-lives (T-DEP) with experimental data (black dots).

Table 4.1: Alpha decay half-lives of $Z = 84 - 118$ isotopes from different formalisms. Q_α experiment is taken from ref. (S. Hosseini et al., 2018), (Bao et al., 2014), (Budaca and Silisteanu, 2011).

Parent Nuclei			Q_α	$logT_{1/2}$(with $T_{1/2}$ in s)				Temperature
Z	A	N	(MeV)	*Exp*	T-IND	T-DEP	UDL	(MeV)
84	188	104	8.069	-3.400	-4.434	-3.979	-4.424	0.899
84	190	106	7.680	-2.600	-3.288	-2.778	-3.229	0.873
84	192	108	7.306	-1.540	-1.748	-1.321	-1.989	0.847
84	194	110	6.974	-0.410	-0.942	-0.416	-0.805	0.824
84	196	112	6.644	0.760	0.241	0.753	0.462	0.800
84	198	114	6.296	2.180	1.653	2.171	1.905	0.776
84	200	116	5.968	3.790	3.157	3.674	3.383	0.752
84	202	118	5.688	5.130	4.472	4.980	4.745	0.731
84	204	120	5.472	6.280	5.592	6.086	5.860	0.714
84	206	122	5.314	7.150	6.608	7.118	6.710	0.700
84	208	124	5.202	7.970	7.203	7.677	7.325	0.680
84	210	126	5.394	7.080	5.884	6.366	6.187	0.699
84	212	128	8.941	-6.520	-6.959	-6.760	-7.318	0.889
84	214	130	7.820	-3.870	-3.737	-3.382	-4.106	0.829
84	216	132	6.893	-0.820	-1.084	-0.852	-0.875	0.776
84	218	134	6.102	2.270	2.181	2.420	2.450	0.727
86	198	112	7.336	-1.190	-1.468	-0.959	-1.282	0.836
86	200	114	7.030	0.000	-0.412	0.102	-0.178	0.814
86	202	116	6.760	1.060	0.631	1.140	0.857	0.795
86	204	118	6.532	2.000	1.89	2.353	1.777	0.778
86	206	120	6.371	2.710	2.16	2.671	2.45	0.765
86	208	122	6.248	3.340	3.319	4.018	2.974	0.754
86	210	124	6.146	3.960	3.185	3.675	3.413	0.744
86	212	126	6.372	3.170	2.061	2.553	2.342	0.754
86	214	128	9.195	-6.570	-6.887	-6.517	-7.223	0.897
86	216	130	8.187	-4.350	-4.285	-4.146	-4.426	0.843
86	218	132	7.249	-1.450	-1.405	-1.150	-1.312	0.791
86	220	134	6.392	1.750	2.651	2.718	2.121	0.741
86	222	136	5.577	5.520	6.148	6.420	6.095	0.690
88	202	114	8.006	-2.580	-2.916	-2.418	-2.749	0.863
88	204	116	7.622	-1.230	-1.702	-1.177	-1.494	0.838
88	206	118	7.402	-0.620	-0.991	-0.480	-0.743	0.822
88	208	120	7.260	0.150	-0.332	0.213	-0.252	0.811
88	210	122	7.139	0.560	-0.013	0.476	0.175	0.800
88	212	124	7.019	1.150	0.317	0.824	0.609	0.790
88	214	126	7.260	0.400	-0.632	-0.128	-0.357	0.799
88	216	128	9.513	-6.740	-7.169	-6.878	-7.286	0.907
88	218	130	8.533	-4.59	-4.812	-4.617	-4.658	0.856
88	220	132	7.579	-1.74	-1.637	-1.442	-1.62	0.805
88	222	134	6.666	1.590	1.566	1.808	1.89	0.752
88	224	136	5.776	5.530	5.747	5.989	6.088	0.699
88	226	138	4.858	10.730	11.307	11.542	11.59	0.639

Table 4.1 continued.

Parent Nuclei			$Q\alpha$	$logT_{1/2}$(with $T_{1/2}$ in s)				Temperature
Z	A	N	(MeV)	*Exp*	T-IND	T-DEP	UDL	(MeV)
90	210	120	8.04	-1.770	-2.403	-1.893	-2.129	0.848
90	212	122	7.939	-1.440	-1.982	-1.45	-1.83	0.839
90	214	124	7.813	-1.000	-1.553	-1.075	-1.436	0.828
90	216	126	8.058	-1.550	-2.354	-1.892	-2.294	0.837
90	218	128	9.836	-7.000	-7.266	-6.881	-7.358	0.918
90	220	130	8.94	-5.010	-5.215	-5.014	-5.046	0.872
90	222	132	8.114	-2.550	-2.81	-2.568	-2.581	0.828
90	224	134	7.285	0.110	0.052	0.307	0.308	0.782
90	226	136	6.438	3.390	3.534	3.785	3.832	0.733
90	228	138	5.507	7.920	8.518	8.774	8.617	0.677
90	230	140	4.757	12.510	13.14	13.361	13.457	0.627
90	232	142	4.069	17.760	18.741	18.955	19.036	0.579
92	218	126	8.773	-3.290	-2.603	-2.755	-3.716	0.868
92	220	128	10.287	-7.220	-7.683	-7.313	-7.724	0.934
92	224	132	8.607	-3.150	-3.568	-3.341	-3.319	0.848
92	226	134	7.688	-0.300	-0.531	-0.266	-0.306	0.799
92	228	136	6.79	2.760	3.358	3.703	3.215	0.749
92	230	138	5.98	6.430	6.684	6.927	7.06	0.701
92	232	140	5.401	9.520	9.902	10.134	10.317	0.664
92	234	142	4.845	13.02	13.655	13.874	13.983	0.627
94	232	138	6.716	4.255	4.034	4.282	4.444	0.739
94	234	140	6.310	5.724	6.28	6.556	6.353	0.713
94	236	142	5.867	7.954	8.245	8.475	8.626	0.686
94	238	144	5.593	9.447	9.958	10.177	10.201	0.667
94	240	146	5.256	11.322	11.968	12.183	12.226	0.645
96	238	142	6.620	4.897	5.482	5.713	5.802	0.724
96	240	144	6.398	6.361	6.688	6.953	6.835	0.709
96	242	146	6.216	7.146	7.265	7.497	7.716	0.696
96	244	148	5.902	8.755	8.956	9.182	9.410	0.676
98	242	144	7.517	2.414	2.352	2.598	2.796	0.764
98	244	146	7.329	3.079	3.322	3.535	3.535	0.751
98	246	148	6.862	5.114	5.086	5.325	5.537	0.724
98	248	150	6.361	7.462	7.551	7.774	7.929	0.695
98	250	152	6.128	8.612	9.073	9.341	9.113	0.680
98	252	154	6.217	7.934	8.114	8.341	8.600	0.682
98	254	156	5.927	9.230	9.666	9.886	10.16	0.664
100	252	152	7.153	4.959	5.512	6.13	5.119	0.730
100	254	154	7.318	4.079	3.927	4.158	4.396	0.735
100	256	156	7.027	5.079	5.172	5.396	5.588	0.718
102	254	152	8.226	1.755	1.314	1.571	1.695	0.779
102	256	154	8.581	0.462	0.079	0.324	0.535	0.790
104	255	151	9.05	0.645	-0.119	0.169	-0.272	0.859
104	256	152	8.93	0.318	0.018	0.337	0.101	0.852
104	258	154	9.19	-0.975	-0.857	-0.534	-0.765	0.860
104	259	155	9.13	0.415	-0.619	-0.287	-0.592	0.856
104	261	157	8.65	0.911	0.925	1.223	0.958	0.830
104	263	159	8.25	3.301	2.321	2.619	2.348	0.808

Table 4.1 continued.

Parent Nuclei			Q_α	$logT_{1/2}$(with $T_{1/2}$ in s)				Temperature
Z	A	N	(MeV)	*Exp*	T-IND	T-DEP	UDL	(MeV)
105	256	151	9.34	0.453	-0.866	-0.545	-0.795	0.871
105	257	152	9.21	0.389	-0.342	0.007	-0.406	0.863
105	258	153	9.50	0.747	-1.341	-0.999	-1.316	0.874
105	259	154	9.62	-0.292	-1.672	-1.363	-1.691	0.878
105	270	165	8.02	3.556	3.504	3.779	3.547	0.787
106	259	153	9.80	-0.507	-1.593	-1.181	-1.827	0.886
106	260	154	9.90	-1.907	-2.238	-1.906	-2.132	0.889
106	261	155	9.71	-0.728	-1.471	-1.181	-1.597	0.879
106	263	157	9.40	0.033	-0.633	-0.338	-0.695	0.861
106	267	161	8.32	2.803	3.049	3.36	2.896	0.805
106	269	163	8.70	2.681	1.434	1.725	1.500	0.820
106	271	165	8.66	2.212	1.541	1.831	1.608	0.815
107	260	153	10.4	-1.456	-2.431	-1.157	-3.136	0.91
107	261	154	10.5	-1.928	-3.534	-3.167	-3.419	0.913
107	265	158	9.38	-0.027	-0.357	-0.043	-0.265	0.857
107	266	159	9.43	0.398	-0.462	-0.16	-0.437	0.858
107	267	160	8.96	1.342	1.36	1.729	1.063	0.835
107	270	163	9.06	1.778	0.606	0.903	0.683	0.835
107	272	165	9.14	0.913	0.693	0.935	0.389	0.835
107	274	167	8.97	1.477	0.855	1.151	0.918	0.825
108	264	156	10.59	-2.796	-3.448	-3.112	-3.336	0.911
108	265	157	10.47	-2.708	-2.937	-2.553	-3.033	0.905
108	266	158	10.35	-2.638	-2.746	-2.432	-2.725	0.898
108	268	160	9.62	0.152	-0.544	-0.193	-0.656	0.863
108	269	161	9.32	1.182	0.44	0.791	0.262	0.848
108	270	162	9.15	0.881	0.719	1.017	0.796	0.839
108	273	165	9.73	-0.041	-1.06	-0.766	-1.068	0.86
108	275	167	9.44	-0.538	-0.165	0.117	-0.213	0.844
109	270	161	10.18	-2.201	-1.529	-1.318	-1.942	0.884
109	274	165	10.04	-0.353	-1.697	-1.387	-1.607	0.871
109	275	166	10.48	-2.013	-2.89	-2.55	-2.851	0.888
109	276	167	9.81	-0.143	-0.85	-0.575	-0.965	0.858
109	278	169	9.59	0.556	-0.166	0.178	-0.329	0.846
110	267	157	11.78	-5	-5.716	-5.349	-5.593	0.955
110	269	159	11.51	-3.747	-5.116	-4.775	-4.998	0.94
110	270	160	11.12	-3.688	-3.882	-3.452	-4.066	0.923
110	271	161	10.87	-2.788	-3.424	-3.056	-3.448	0.911
110	273	163	11.37	-3.77	-4.74	-4.372	-4.728	0.928
110	277	167	10.83	-2.387	-3.511	-3.172	-3.442	0.899
110	279	169	9.84	0.301	-0.72	-0.432	-0.713	0.855
110	281	171	8.86	2.301	2.371	2.652	2.432	0.809
111	274	163	11.362	-2.194	-4.286	-3.993	-4.373	0.926
111	279	168	10.568	-0.77	-2.104	-1.663	-2.411	0.885
111	280	169	9.938	0.556	-0.573	-0.295	-0.634	0.858
111	282	171	9.177	-0.292	1.794	2.094	1.742	0.822

Table 4.1 continued.

Parent Nuclei			Q_α	$logT_{1/2}$(with $T_{1/2}$ in s)				Temperature
Z	A	N	(MeV)	*Exp*	T-IND	T-DEP	UDL	(MeV)
112	277	165	11.666	-3.155	-4.811	-4.5	-4.803	0.933
112	283	171	9.725	0.58	0.495	0.755	0.357	0.844
112	285	173	9.328	1.462	1.514	1.793	1.598	0.824
113	278	165	11.899	-3.62	-5.095	-4.738	-5.02	0.94
113	283	170	10.313	-1	-1.102	-0.803	-1.013	0.869
113	284	171	10.191	-0.319	-0.662	-0.339	-0.676	0.862
113	285	172	9.927	0.738	0.048	0.352	0.095	0.849
113	286	173	9.815	1.293	1.245	1.28	0.423	0.843
114	286	172	10.373	-0.886	-0.943	-0.648	-0.856	0.866
114	287	173	10.211	-0.319	-0.169	0.197	-0.401	0.858
114	288	174	10.08	-0.097	-0.091	0.193	-0.027	0.851
114	289	175	10.007	0.415	0.072	0.363	0.177	0.847
115	287	172	10.789	-1.456	-1.6	-1.322	-1.664	0.882
115	288	173	10.677	-1.06	-1.49	-1.183	-1.372	0.876
115	289	174	10.504	-0.658	-0.702	-0.454	-0.902	0.867
115	290	175	10.139	-1.796	0.069	0.352	0.148	0.851
116	290	174	11.042	-2.167	-1.844	-2.03	-2.152	0.841
116	291	175	10.94	-1.699	-1.414	-1.772	-1.748	0.835
116	292	176	10.858	-1.397	-0.938	-1.565	-1.114	0.831
116	293	177	10.736	-1.097	-1.063	-1.245	-1.36	0.825
117	293	176	11.233	-1.824	-1.123	-2.221	-0.976	0.843
117	294	177	11.01	-1.114	-1.334	-1.645	-1.659	0.834
117	293	176	11.03	-1.678	-1.396	-1.683	-1.719	0.836
118	294	176	11.862	-3.046	-2.483	-3.473	-1.904	0.865
118	295	177	11.76	-3	-3.026	-3.239	-3.331	0.86

Table 4.2: Root mean square error associated with predicted half-lives

Model/Formalism	Isotopes	RMSE (σ)
Cubic Plus Proximity without Temp.	HN	0.51
Cubic Plus Proximity with Temp.	HN	0.47
Universal decay law (UDL)	HN	0.53
Coulomb Plus Proximity model (CPPM) (S. Hosseini et al., 2018)	HN	0.61
Cubic Plus Proximity without Temp.	SHN	0.88
Cubic Plus Proximity with Temp.	SHN	0.72
Universal decay law (UDL)	SHN	0.89
Cubic Plus Proximity without Temp.	HN + SHN	0.72
Cubic Plus Proximity with Temp.	HN + SHN	0.60
Universal decay law (UDL)	HN + SHN	0.73

4.4 Deformation dependent decay life-times

Nuclear deformation occurs when excess nucleons are present in an open-shell configuration. They also occur due to the influence of rotation on nuclei and the excited states. The atomic nuclei, as is well known, are charged particles that interact with each other via nuclear and electromagnetic forces. But, the interactions between a pair of spherical - spherical nuclei, spherical - deformed nuclei, and deformed - deformed nuclei, on the other hand, are distinct. Spherical symmetry in nuclear shape is due to the influences of its internal structure. Any deviation from spherical symmetry, in general, can be expressed in terms of the deformation parameter, which is a measure of the shape deformation.

A decay process involving a spherical daughter and cluster emission are not orientation and direction-dependent, whereas either/both the daughter and fragment nuclei being deformed, will in general be orientation and direction dependent. To understand the deformation effects in detail, parameters associated with fusion reactions and decay mechanism, like the touching distance, barrier height, formation probability in fusion and tunneling probability should be studied as they can certainly relate to the extent to which the nuclei are deformed and in a way

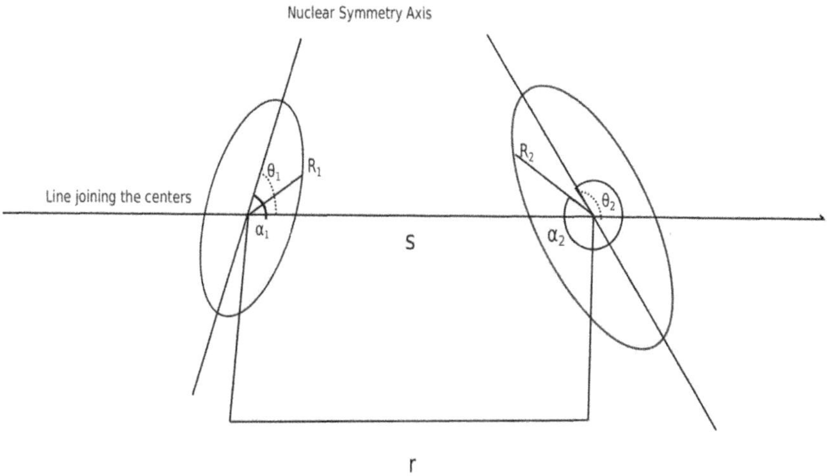

Figure 4.8: Deformed and oriented nuclei lying on the same plane.

gather information about what are their orientations. In this context, a brief description on the various nuclear properties and potentials that get modified for alpha/cluster decay in deformed systems is discussed based on the ideas gathered from ref. (Malhotra and Gupta, 1985; Gupta, Balasubramaniam, Kumar, Singh, Manhas and Greiner, 2005).

Consider a fragment nuclei moving with respective to daughter, which is schematically represented in fig. 4.8, where r is the distance between the daughter and cluster nuclei and θ_1 and θ_2 are the orientation angles. From fig. 4.8, θ_i is the angle between the polar axis and the line connecting the center of nuclei, measured in anticlockwise direction. The angle α_i is measured clockwise from the nuclear symmetry axis. The effective radius of this deformed nuclei along any arbitrary direction is given as,

$$R_i(\alpha_i) = R_{0i}\left[1 + \sum_{\lambda=2,4,6} \beta_{\lambda i}\, Y_\lambda^0(\alpha_i)\right], \tag{4.90}$$

where β's are deformation parameters obtained using FRDM (Moller et al., 2016) and $Y_\lambda^0(\alpha_i)$ are the spherical harmonics. The nuclear radii along a given direction in terms of deformation parameters, are corrections to the spherical radii as discussed in chapter 2. The generalized deformation dependent nuclear proximity potential formula is of the form,

$$V_n(r, \theta_1, \theta_2) = 4\pi\gamma b F(\theta)\phi(s), \tag{4.91}$$

where F is the mean curvature that depends on shape and geometry of the deformed nuclei

$$F = \left(\frac{R_{11}R_{12}R_{21}R_{22}}{R_{21}R_{22} + R_{11}R_{12} + R_{21}R_{12} + R_{11}R_{22}} \right)^{1/2}. \tag{4.92}$$

$\phi(s)$ depends only on the shortest distance between the nuclei, ie., the shortest distance between the nuclear surfaces 's'. γ can be calculated using eqn. 2.46 and $\phi(s)$ can be determined using eqn. 2.47 and 2.48 discussed in chapter 2 once s is estimated.

The four principal radii of curvature as defined in eqn. 4.92 are given by,

$$R_{i1}(\alpha_i) = \frac{\left[R_i^2(\alpha_i) + R_i'^2(\alpha_i)\right]^{3/2}}{\left[R_i^2(\alpha_i) + 2R_i'^2(\alpha_i) - R_i'(\alpha_i)R_i''(\alpha_i)\right]}; \quad R_{i2}(\alpha_i) = \frac{R_i(\alpha_i)\sin(\alpha_i)}{cos(\frac{\pi}{2} - \alpha_i - \delta_i)}, \tag{4.93}$$

where $\delta_1 = \theta_1 - \alpha_1$ and $\delta_2 = \pi + \theta_2 - \alpha_2$. For the given orientation angles θ_1 and θ_2, the shortest distance between nuclear surface s in femtometer, is given as,

$$s = r - R_1(\alpha_1)\cos(\theta_1 - \alpha_1) - R_2(\alpha_2)\cos(\pi + \theta_2 - \alpha_2). \tag{4.94}$$

From fig. 4.8 it is clear that s is the distance between the center of daughter and cluster nuclei subtracted by the projection of daughter and cluster nuclei along the line joining of the nuclei. To estimate s, α_i's are required, which can be obtained by the condition that s has to be minimum. That is,

$$\frac{ds}{d\alpha_1} = \frac{ds}{d\alpha_2} = 0. \tag{4.95}$$

The simplified minimization conditions are given (Gupta, Balasubramaniam, Kumar, Singh, Manhas and Greiner, 2005; Malhotra and Gupta, 1985) as,

$$tan(\theta_1 - \alpha_1) = -\frac{R_1'(\alpha_1)}{R_1(\alpha_1)}; \quad tan(\pi + \theta_2 - \alpha_2) = -\frac{R_2'(\alpha_2)}{R_2(\alpha_2)}. \tag{4.96}$$

On solving numerically, the angle α_i corresponding to a given orientation (θ_i) can be determined and hence the shortest distance between two surfaces of nuclei can be found out, which governs the nuclear interaction. Coulomb potential between non-spherical shaped objects can be expressed as Coulomb potential due to spherical shapes plus terms associated with deformation. In similar lines, the Coulomb potential for deformed daughter and cluster is given as (Gupta, Balasubramaniam, Kumar, Singh, Manhas and Greiner, 2005; Pahlvani et al., 2013; Wong, 1973),

$$V_c(r,\theta_1,\theta_2) = \frac{Z_1Z_2e^2}{r} + 3Z_1Z_2e^2 \sum_{\lambda,i=1,2} \frac{1}{2\lambda+1} \frac{R_i^{\lambda}(\alpha_i)}{r^{\lambda+1}} Y_{\lambda}^{0}(\theta_i)[\beta_{\lambda i} + \frac{4}{7}\beta_{\lambda i}^{2} Y_{\lambda}^{0}(\theta_i)]. \tag{4.97}$$

Thus the modified potentials that are deformation dependent when incorporated to the interaction potential in exterior region will be,

$$V_{ext}(r,\theta_1,\theta_2) = V_n(r,\theta_1,\theta_2) + V_c(r,\theta_1,\theta_2) + V_l(r). \tag{4.98}$$

Here $V_l(r)$ is considered as zero, as the nuclei are treated in ground state condition ie., l= 0. Deformation dependence in the post-scission region of nuclear radii R and orientation angles, is same for the pre-scission region. A deformed parent nucleus is an ellipsoid, its longer radius is known as equatorial radius or semi-major axis, and the shorter radius is known as polar radius or semi-minor axis. The assumption is made that the fission direction is along the major axis of the parent nuclei. r_i is the distance between the center of mass location of the daughter and the cluster inside the deformed parent nuclei and can be obtained using the planar section cut method (Umai Parvathiy, 2016) as,

$$h_1 + h_2 = 2a_0 \tag{4.99}$$

where h_1 and h_2 are the heights of deforming daughter and cluster nuclei and a_0 is the semi major axis of deformed parent nuclei. Then,

$$\frac{3h_1}{a_0} + \left(1 - \frac{h_1}{a_0}\right)^3 - 1 = 4\frac{A_1}{A}. \tag{4.100}$$

Hence, h_1 can be obtained from eqn. 4.100 and is used to find r_i as,

$$r_i = \frac{3}{4}\left[\frac{h_1^2}{a_0+h_1} + \frac{h_2^2}{a_0+h_2}\right], \tag{4.101}$$

where $a_0 = R_{00}\left[1+\beta_{p,2}\,Y_2(\alpha)+\beta_{p,4}\,Y_4(\alpha)+\beta_{p,6}\,Y_6(\alpha)\right]$, with $\alpha = 0$ for a prolate nucleus and $\alpha = \pi/2$ for an oblate nuclei for major axis. The shape deformation of parent, daughter and cluster nuclei modifies the r_i, r_t, $V_{ext}(r_t)$ which in turn results in the modified pre-scission potential, such as,

$$V_{ov}(r,\theta_1,\theta_2) = (-E_v+Q) + [V_{ext}(rt)+E_v-Q]\left[s_1\left(\frac{r-r_i}{r_t-r_i}\right)^2 - s_2\left(\frac{r-r_i}{r_t-r_i}\right)^3\right]. \tag{4.102}$$

The overlapping region $V_{ov}(r,\theta_1,\theta_2)$ can be determined using E_v, s_1 and s_2 as discussed in chapter 2.

4.5 Particle and clusters emissions from deformed emitters

The Nuclear radii in deformed systems are direction-dependent based on the orientation angle, given in eqn. 4.90, which is proportional to the spherical radii R_{0i} and is evident that in heavy and superheavy nuclei the deformation effects will be significant as they possess larger radii. Furthermore, the touching distance, which is simply the sum of projected nuclear radii along the line joining the centers of nuclei, experiences the impact of the nuclear radii changes in response to changes in orientation angle between daughter and cluster, and this directly influences the interaction potential that governs decay half-lives.

Figure 4.9 details the interaction potential for the alpha decay process from the superheavy nuclei $^{294}118$ as deformation dependent (β-DEP) and spherical case ($\beta - IND$), and also for different orientation angles $(\theta_1 - \theta_2)$ of the daughter nuclei and cluster nuclei. It is evident from the figure that the interaction potential for the spherical case is higher than the $\beta - DEP$ case with 0° – SPH or 180° – SPH orientation, which suggests that the inclusion of deformation has reduced the barrier height and width by some proportion. It is also noticed that when $V(r) = Q$ the turning points are seen to get modified. With these findings it is possible to ensure that the

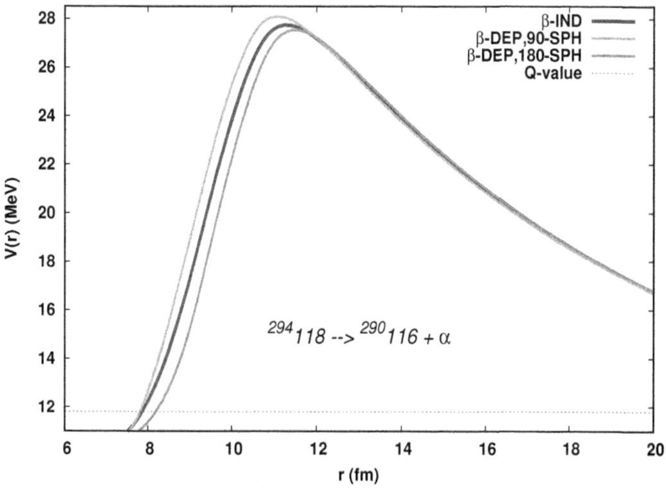

Figure 4.9: Interaction potential of decay process $^{294}118 \rightarrow {}^{290}116 + \alpha$, with different orientation angles and with/without deformation parameters.

tunneling probability gets increased and thereby the lowering of decay half-life.

The alpha particle is said to be spherical in nature, so the changes in the barrier height and width seen are primarily due to the parent and daughter deformations. The $90^\circ - \mathrm{SPH}$ orientation with $\beta - DEP$ case, as seen in fig. 4.9, leads to an increase in barrier height and thus suppressing the tunneling chances. In $^{294}118 \rightarrow {}^{290}116 + \alpha$, the configuration where the prolate daughter ($\beta_2 = 0.064$) is at orientation angle $\theta_1 = 90^\circ$ alongside spherical alpha particle as represented in fig. 4.10, is a compact configuration having the smallest interaction radius and largest barrier, on the other hand, the configuration where the prolate daughter's orientation is at $\theta_1 = 180^\circ$ (or 0°) is an elongated one having long interaction radius and smallest barrier. For any other orientations angle apart from $\theta_1 = 90^\circ$ & $\theta_2 = 180^\circ$, the barrier will be at its intermediate form between the two. These two configurations are called optimum configurations, the compact one is associated with hot fusion reaction, and the long and elongated one is associated with cold fusion reaction. For oblate/prolate/spherical combinations of daughter and cluster, the optimum configurations are shown in fig. 4.10.

The interaction potential for cluster decay ^{28}Mg from the ^{235}U nucleus is represented in fig. 4.11, considering the parent ($\beta_2 = 0.215$), daughter ($\beta_2 = -0.011$) and cluster($\beta_2 = 0.277$) are

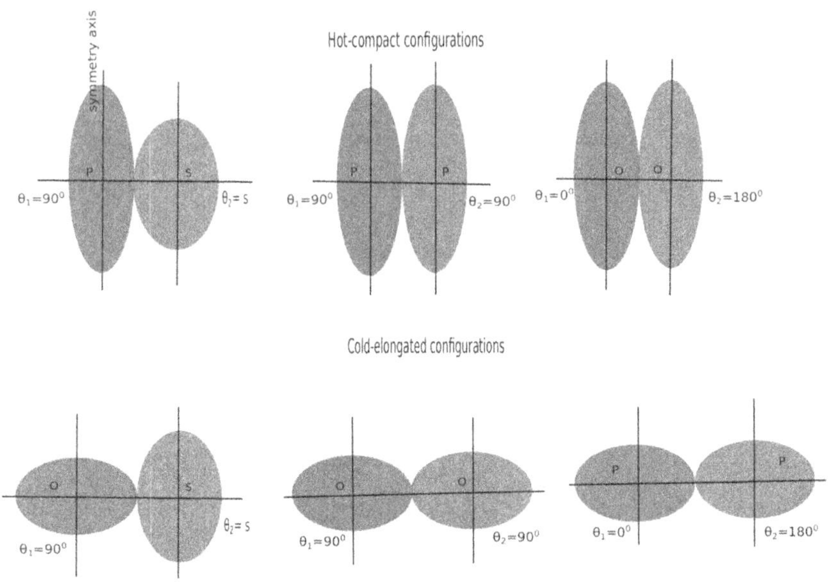

Figure 4.10: The optimum orientations associated with nuclear fusion and decay process. S denotes spherical, o denotes oblate and p denotes prolate nuclei.

deformed. The interaction potential for the deformed parent, deformed parent, and daughter, and the case where all three nuclei are deformed is presented at orientation angle $\theta_1 = 90^\circ$ and $\theta_2 = 90^\circ$ with hexacontatetrapole deformations. This is an equator-equator orientation. It is evident from fig. 4.11, that the interaction potential with the deformed parent becomes smaller than that of the spherical one, and there is a significant change in the turning point but the height of the barrier remains constant as it is mainly governed by the daughter and cluster. The inclusion of deformation of the daughter doesn't have a major effect on the barrier since $\beta_{4,6} = 0$ for the daughter involved in this decay process, but the inclusion of cluster deformation to existing parent and daughter leads to considerable change in barrier height, making the barrier width smaller and barrier height higher. The interaction potential shown for the orientation $\theta_1 = 90^\circ, \theta_2 = 180^\circ$, is the cold optimum configuration where the barrier is significantly smaller. On excluding the $\beta_{4,6}$ parameter of the cluster, it is observed that these parameters contribute toward barrier elongation. These two cases are also presented in fig. 4.11. Hence nuclear deformation and orientations are important factors influencing the stability of nuclei.

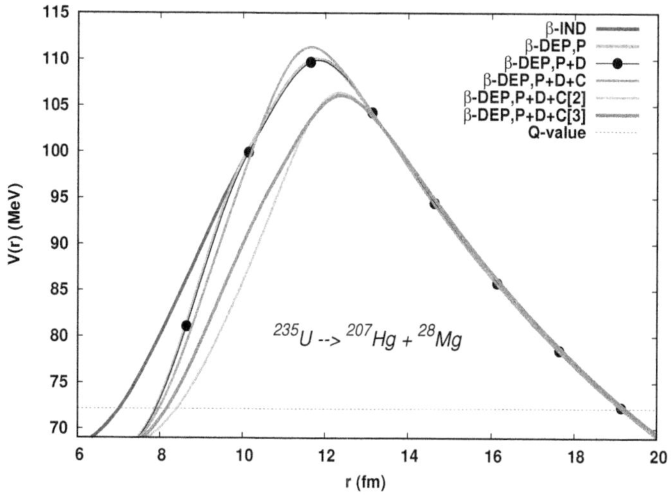

Figure 4.11: Interaction potential of decay process $^{235}U \rightarrow^{207} Hg +^{28} Mg$ at orientation angle $90^\circ - 90^\circ$ with/without deformation parameters of parent, daughter and cluster. Additionally the interaction potential is shown for two cases : [2] is for all deformations with orientation angle $90^\circ - 180^\circ$, [3] is for $\beta_{4,6}$=0 of cluster.

The tunneling probability through the pre-scission region described by cubic potential can be interpreted as preformation probability, that is the probability of formation of the cluster within the parent's surface before tunneling out. The parent nuclei being spherical or deformed can have considerable influence to pre-scission preformation probability. In this work the clusters preformation probability for ^{14}C, ^{22}Ne, ^{28}Mg and ^{34}Si from heavy isotopes under different conditions are obtained, ie., one with no-deformation of parent nuclei, then by incorporating the deformation parameters (β_i's) from lower to the higher order are presented in fig 4.12. The deformed parent enhances the preformation probability of the cluster, and is seen to change by a certain order with each higher-order deformation parameter taken into account. For several nuclei the β_6 is considered as zero, and so taking this parameter into account has not altered the preformation factor, but for those nuclei where β_6 is non-zero, ie., the positive β_6 values contributes towards increasing the preformation probability and the negative β_6 is seen to contribute towards decreasing the preformation probability, which is seen for ^{28}Mg emission from ^{236}U and ^{34}Si emission from ^{240}Pu,^{241}Am,^{242}Cm. The lighter cluster's preformation probability is seen to be more than that of heavier ones.

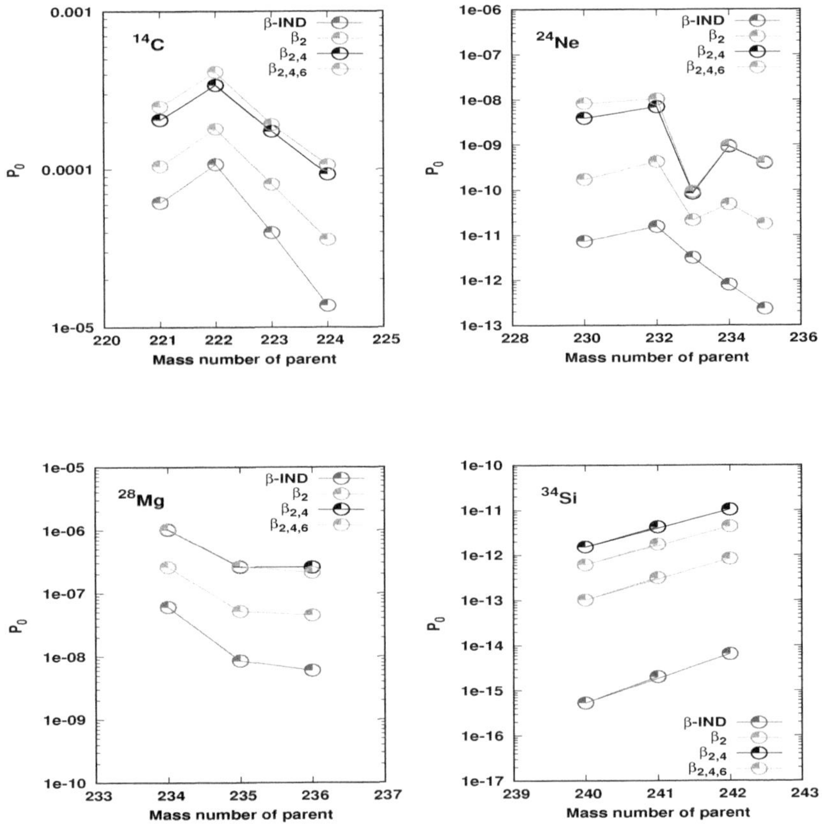

Figure 4.12: Preformation probability for clusters from spherical and deformed parent is shown, the parent's deformation parameters are plotted under β_2, $\beta_{2,4}$, and $\beta_{2,4,6}$ condition.

The half-lives are obtained at the cold optimum configuration for all decays, and for the ones involving prolate clusters intermediate orientation configurations are employed ($\theta_2 = 90°$). The decay half-lives for (i) the spherical case (all deformations zero), (ii) with only parent is deformed (P:$\beta_{2,4,6}$), (iii) with both parent and daughter are deformed (P+D:$\beta_{2,4,6}$), and (iv) with parent, daughter, and cluster are deformed (P+D+C:$\beta_{2,4,6}$), is presented in table 4.3. The half-life values are calculated by using up to hexacontatetrapole β_6, hexadecapole β_4 and quadrupole deformations β_2, to bring out the role of deformation parameters. The decay lifetimes of some clusters are shown in fig. 4.13 and 4.14 for comparison. In fig. 4.13, ^{14}C emission from $^{221-226}Ra$ parent shows a decrease in half-life value when parent deformation upto β_6 is considered as compared to spherical case. Further inclusion of parent and daughter deformation upto β_6 doesn't show considerable change in the half-life value as the daughter and cluster have βs to be zero. The exclusion of β_6 deformation parameter from parent, daughter, and cluster results in an increase of half-life. Half-life further increases on excluding $\beta_{4,6}$ deformation parameters and considering only quadruple deformation. From ^{14}C emission the spherical, P+D+C:$\beta_{2,4,6}$ taken up to hexacontatetrapole and P+D+C:β_2 quadrupole deformation is close to the experimental values.

Similarly, being a deformed cluster the half-life plots of ^{24}Ne bring out different details. From the plot, it is seen that the cluster deformation when included in the existing parent and daughter deformation, causes considerable change in half-life. The influence of daughter deformation is also seen in the decay process associated with ^{233}U. For the isotope of Uranium, spherical and P+D+C:$\beta_{2,4,6}$ taken up to hexacontatetrapole is close to the experimental value. The half-life plots of ^{14}C, ^{24}Ne, ^{30}Mg and ^{34}Si from fig. 4.13 and 4.14 suggest the inclusion of parent and daughter deformation lowers the half-life as compared to the spherical case, which is in accordance with other studies (Santhosh and Joseph, 2002b). It is mainly the cluster deformation that results in the larger barrier and increase in half-life. Without the deformation of the cluster, the calculated half-life values will not be in close correlation with the experimental values. Hence cluster deformations are important for the Cubic plus proximity model. In terms of deformation parameters, we notice that the inclusion of quadrupole deformation (β_2) on the parent, daughter, and cluster lowers the half-life as compared to the spherical case, and taking up to hexadecapole ($\beta_{2,4}$) further lowers the half-life. This is in consistent with the studies carried out by (Santhosh et al., 2009). Most of the studies dealt with alpha and cluster decay have not considered the β_6 parameter. It is found that the addition of β_6 increases the barrier height and increases half-life for non-spherical clusters, and contributes towards lowering the half-life for spherical cluster ^{14}C. (Pahlavani et al., 2013) studied the role of daughter deformation parameter β_6 for alpha decay half-lives of superheavy nuclei and found it to contribute towards decreasing the potential barrier particularly at near the barrier distance and established its importance. The present study considers hexacontatetrapole deformation of the parent, daughter, and cluster, and establishes the role of this parameter further, and also for deformed clusters.

The ^{28}Mg decay shows a lower half-life on the inclusion of cluster deformation compared to the parent and daughter deformation conditions (fig. 4.14). The deformation parameters of this cluster are ($\beta_2 = 0.277, \beta_4 = -0.073$, $\beta_6 = 0.008$), and the hexadecapole deformation value is larger for ^{28}Mg compared to other clusters considered in this study, and this parameter may play a crucial role in the lowering the half-life. Same is obtained for ^{20}O half-lives presented in table 4.3 whose deformation parameters are ($\beta_2 = 0.010$, $\beta_4 = -0.024$, $\beta_6 = 0.020$).

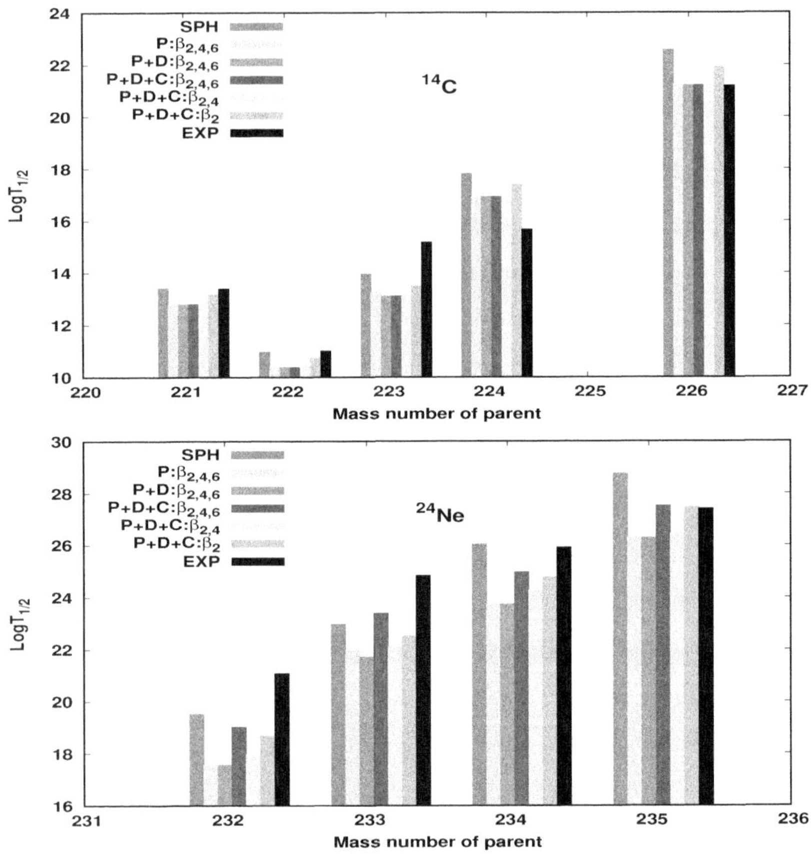

Figure 4.13: Spherical and Deformation-dependent half-lives of clusters emitted against the mass number of Parent nuclei in the heavy region. The heights of the histogram represent the decay half-life values.

In this range of trans-actinide elements, for cluster decay, the half-lives of spherical nuclei are higher than the deformed ones for most of the decay processes. The calculated half-lives for cluster decay compared with experimental half-lives are in good agreement with each other and also with the half-lives estimated from other theoretical formalisms. The RMS error for the deformation dependent model is calculated using eqn. 4.89, The obtained values, $\sigma(\beta-IND) = 1.83$, $\sigma(\beta^{P}_{2,4,6}) = 2.70$, $\sigma(\beta^{P+D}_{2,4,6}) = 2.71$, $\sigma(\beta^{P+D+C}_{2,4,6}) = 2.31$, $\sigma(\beta^{P+D+C}_{2,4}) = 3.94$, $\sigma(\beta^{P+D+C}_{2}) = 2.50$, suggests that the deformation-dependent model involving parent, daughter, and cluster deformation up to hexacontatetrapole performs better than other deformation-dependent models and has better predictive power. Next, cluster emission from the superheavy nuclei is analyzed, due to their size the deformations may significantly alter the half-life. The decay half-lives of 30 different light clusters in the range $Z = 2-20$ and mass number $A = 4-46$ are studied; of which all isotopes are stable except ^{14}C and ^{20}O. The emission of these clusters from superheavy isotopes $^{298,300}118$, $^{300,312}120$, $^{310,312}122$, and $^{312,314}124$ are being calculated. The decay half-life versus the mass number of clusters is presented in fig. 4.15 - 4.18. The dips in the half-life curve suggest that emission of certain clusters from the parent nuclei occurs rapidly and are relatively favorable compared to the nearby clusters. The inclusion of β_2 deformation results in lowering the half-life but the trend of dips is the same as that of the spherical case. The inclusion of up to β_6 deformation results in new dips. The ^{20}O cluster decay seems to be favorable due to the deformation effects. The deformation effects also favor ^{33}Si cluster compared to ^{34}Si. In general, the deformation effect enhances or lowers the half-life and is seen clearly in fig. 4.15 - 4.18. For some decay processes, the deformation enhances the decay whereas in some suppresses it. The deformation dependent ($\beta_{2,4,6}$) decrease of decay half-life for clusters with $A = 25-30$ in $^{300}120$ involves a prolate daughter and oblate cluster, whereas the deformation dependent increase in half-life seen for clusters with $A = 20-25$ involves a prolate daughter and prolate cluster. The decrease in half-life for all the cluster emissions is obtained from $^{314}124$ parent which is prolate in shape, whereas $^{298,300}118$, $^{300,312}120$, $^{310,312}122$, $^{312}124$ are oblate/spherical, which suggests the prolate shape of parent governing the decay. This analysis brings out the shape deformation enhanced and suppressed cluster emissions. The experimental observation limits, $10^{-6} \leq T_{1/2} \leq 10^{30}$ s are also indicated in figs. 4.15 - 4.18, which indicates the clusters which may be observed in experiments might compete with the alpha decay.

In the heavy region the cluster decays which lead to the formation of the doubly magic nuclei

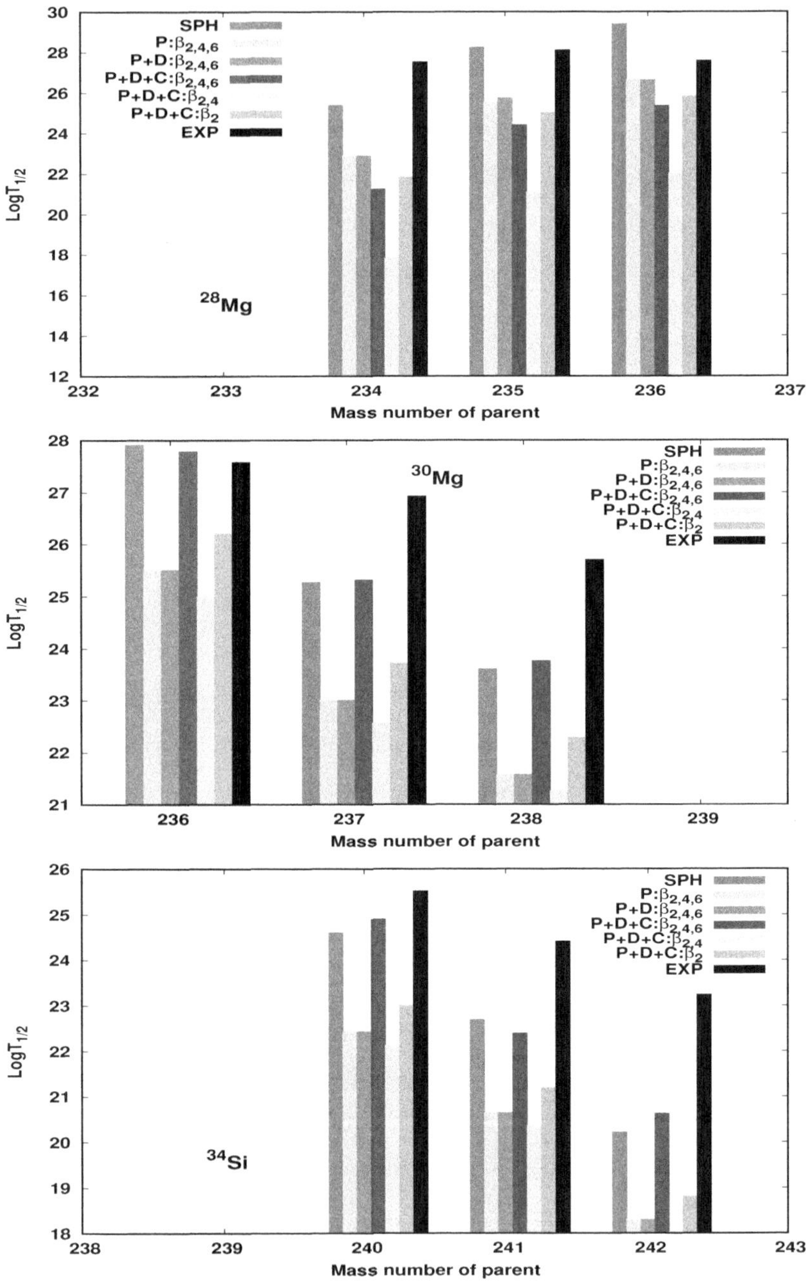

Figure 4.14: Same as in fig. 4.13 but for $^{28,30}Mg$ and ^{34}Si

^{208}Pb are most favourable ones. In the superheavy region, Z=114 and N=184 nuclei are expected to be doubly magic. The formation of the doubly magic daughter $^{298}114$ from different decaying parents with emission of clusters will be the most favorable decay chain (fig. 4.19 and table 4.4). The ^{4}He emission is expected to be more favorable from Z=116 than from any other parent, ^{14}C is more favorable from Z=122 isotope than from other parents, and similar is the case for bigger clusters. These are the possible cluster decays that are favorable in forming doubly magic nuclei from their respective parent. The branching ratio is the ratio of the fraction of nuclei that will decay via cluster emission concerning nuclei decaying via alpha emissions, and it can be obtained using, $BR = \frac{T_{1/2}^{\alpha}}{T_{1/2}^{cluster}}$. Among the lighter clusters, ^{14}C, ^{20}O and ^{26}Mg exhibit lower decay half-lives, and among the medium clusters, $^{33,34}Si$ and ^{40}Ar exhibit lower half-life values. The branching ratio is higher for these clusters. Depending on the branching ratio, some of these rare cluster decays may occur and can be identified during the synthesis or decay-related experiments in near future. It was reported that in heavy nuclei, the experiments have been able to detect cluster emissions occurring with observed branching ratios as low as $10^{-9} - 10^{-17}$ (Soylu and Qi, 2021).

Table 4.3: Comparison of calculated logarithmic cluster decay half-lives from heavy isotopes with/without deformation with that of experimental values, Universal decay law (UDL) and Horoi empirical formulae.

Parent		Cluster		Q	$logT_{1/2}$(with $T_{1/2}$ in s)								
A	Z	A_2	Z_2	(MeV)	SPH	$\beta^{P}_{2,4,6}$	$\beta^{P+D}_{2,4,6}$	$\beta^{P+D+C}_{2,4,6}$	$\beta^{P+D+C}_{2,4}$	β^{P+D+C}_{2}	Exp	UDL	Horoi
221	88	14	6	32.39	13.42	12.81	12.81	12.81	12.90	13.19	13.39	14.56	12.28
222	88	14	6	33.05	10.97	10.39	10.39	10.39	10.47	10.75	11.01	13.15	11.00
223	88	14	6	31.85	13.96	13.26	13.11	13.11	13.15	13.49	15.19	15.60	13.38
224	88	14	6	30.53	17.82	16.93	16.93	16.93	16.99	17.40	15.68	18.47	16.15
226	88	14	6	28.21	22.57	21.22	21.22	21.22	21.30	21.92	21.19	23.98	21.50
226	90	14	6	30.67	18.79	17.71	17.71	17.71	17.78	18.26	15.30	20.29	17.83
221	87	14	6	31.29	13.80	13.15	13.15	13.15	13.24	13.55	14.52	15.81	13.54
225	89	14	6	30.48	18.05	17.08	17.08	17.08	17.14	17.59	17.16	19.66	17.26
228	90	20	8	44.72	21.33	19.94	19.94	17.51	19.40	20.64	20.72	23.57	21.20
230	92	22	10	61.59	20.89	18.38	18.38	21.38	20.69	15.20	18.20	21.07	19.28
231	91	23	9	51.84	23.89	22.08	22.08	25.04	23.67	22.85	26.02	25.84	23.85
230	92	24	10	61.55	21.21	19.05	19.05	20.67	19.41	20.20	18.20	22.40	20.99
232	92	24	10	62.31	19.53	17.57	17.57	19.02	17.96	18.69	21.08	20.97	19.94
233	92	24	10	60.50	22.98	21.96	21.71	23.40	22.12	22.53	24.83	23.99	22.53
234	92	24	10	58.84	26.04	23.74	23.74	24.97	24.28	24.77	25.92	26.87	25.01
235	92	24	10	57.36	28.77	26.29	26.29	27.53	26.47	27.48	27.42	29.54	27.31
230	90	24	10	57.78	24.71	22.40	22.40	23.64	23.09	24.39	24.61	25.83	23.88
232	90	24	10	55.62	28.89	26.47	26.47	27.71	26.77	27.66	29.20	29.73	27.32
231	91	24	10	60.42	21.53	19.52	19.52	20.71	20.48	20.58	23.23	22.69	21.30
232	90	26	10	55.97	28.99	26.93	26.93	27.58	23.49	27.54	29.20	30.47	28.56
234	92	26	10	59.47	25.02	23.13	23.13	24.22	19.92	24.64	25.92	27.02	25.75
236	92	26	10	56.75	30.43	28.19	28.19	26.92	23.46	27.49	25.90	32.10	30.17
234	92	28	12	74.13	25.40	22.90	22.90	21.26	17.83	21.84	27.54	25.03	24.74
235	92	28	12	72.20	28.26	25.53	25.76	24.42	21.10	25.02	28.10	28.12	27.28
236	92	28	12	71.69	29.41	26.64	26.64	25.38	22.02	25.84	27.58	28.89	27.98
236	94	28	12	79.67	19.56	17.41	17.41	16.16	12.65	16.46	21.67	19.83	20.83
237	93	30	12	75.02	25.27	23.01	23.01	25.32	22.57	23.72	26.93	26.12	26.51
236	92	30	12	72.51	27.92	25.51	25.51	27.79	25.01	26.21	27.58	28.56	28.36
238	94	30	12	77.03	23.60	21.58	21.58	23.77	21.27	22.29	25.70	24.59	25.41
241	95	34	14	93.84	22.69	20.65	20.65	22.40	20.37	21.19	24.41	22.10	25.57
240	94	34	14	90.95	24.61	22.43	22.43	24.90	22.13	23.00	25.52	24.47	27.22
242	96	34	14	96.53	20.22	18.30	18.30	20.63	18.03	18.81	23.24	20.10	24.22

Table 4.4: Alpha and cluster decay of SHN leading to the formation of doubly magic daughter nucleus $^{298}114$.

Parent		Cluster		Daughter		Q	$logT_{1/2}$(with $T_{1/2}$ in s)					Branching ratio	
A	Z	A_2	Z_2	A_1	Z_1	(MeV)	$\beta-IND$	$\beta-DEP$	UDL	Horoi	α	$\beta-IND$	$\beta-DEP$
302	116	4	2	298	114	12.08	-3.84	-3.84	-4.79	-4.06	-3.84	1.00E+00	1.00E+00
304	117	6	3	298	114	9.81	34.78	34.72	35.06	31.20	-5.88	2.17E-41	2.50E-41
307	118	9	4	298	114	22.67	18.99	20.35	20.01	18.94	-5.16	7.15E-25	3.13E-26
308	118	10	4	298	114	23.47	17.99	18.85	19.86	19.61	-3.93	1.21E-22	1.66E-23
310	120	12	6	298	114	43.97	16.56	18.77	16.64	15.90	-1.07	2.33E-18	1.45E-20
312	120	14	6	298	114	46.19	13.76	15.13	15.16	15.88	-0.96	1.89E-15	8.06E-17
314	122	16	8	298	114	64.12	20.44	22.60	19.81	20.37	-4.71	7.08E-26	4.86E-28
316	122	18	8	298	114	66.63	17.21	22.77	17.73	20.00	-4.37	2.60E-22	7.15E-28
318	122	20	8	298	114	66.74	17.46	13.66	18.95	22.10	-3.24	1.97E-21	1.24E-17
318	124	20	10	298	114	81.94	28.28	35.07	26.33	27.83	-6.15	3.69E-35	6.00E-42
320	124	22	10	298	114	87.36	20.82	28.02	20.06	24.57	-5.52	4.66E-27	2.91E-34
322	124	24	10	298	114	90.47	16.39	23.63	16.78	23.38	-4.40	1.61E-21	9.33E-29
324	124	26	10	298	114	88.60	19.01	22.85	20.19	27.08	-1.79	1.57E-21	2.31E-25
322	126	24	12	298	114	97.22	39.37	29.10	35.99	37.53	15.21	6.98E-25	1.31E-14
323	126	25	12	298	114	98.88	37.35	27.81	34.28	37.06	18.16	6.38E-20	2.25E-10
324	126	26	12	298	114	104.12	30.04	20.03	27.52	33.07	19.56	3.31E-11	3.38E-01
326	126	28	12	298	114	112.05	19.07	18.42	17.91	27.69	-2.31	4.14E-22	1.85E-21
326	128	28	14	298	114	118.59	41.43	28.27	36.61	41.12	-5.56	1.02E-47	1.49E-34
327	128	29	14	298	114	120.31	39.30	26.68	34.81	40.64	-5.37	2.11E-45	8.93E-33
328	128	30	14	298	114	125.49	32.46	22.52	28.44	37.14	-5.48	1.17E-38	1.01E-28
329	129	31	15	298	114	132.16	39.07	25.07	33.61	41.61	-5.35	3.83E-45	3.81E-31
330	130	32	16	298	114	144.54	37.87	34.52	31.51	41.39	-14.91	1.67E-53	3.68E-50
331	130	33	16	298	114	146.99	35.27	18.12	28.83	40.42	-15.01	5.17E-51	7.34E-34
332	130	34	16	298	114	151.79	28.73	22.40	23.36	37.66	-14.39	7.58E-44	1.61E-37
333	131	35	17	298	114	158.08	35.50	27.76	28.74	42.29	-15.53	9.37E-52	5.16E-44
334	132	36	18	298	114	165.55	40.78	32.29	32.66	45.92	-5.81	2.57E-47	8.07E-39
336	132	38	18	298	114	174.65	29.55	25.37	22.52	41.18	-6.57	7.48E-37	1.13E-32
338	132	40	18	298	114	180.35	21.96	20.81	16.28	38.77	-6.12	8.25E-29	1.19E-27
338	134	40	20	298	114	188.67	41.67	36.55	31.18	49.01	-6.91	2.65E-49	3.48E-44

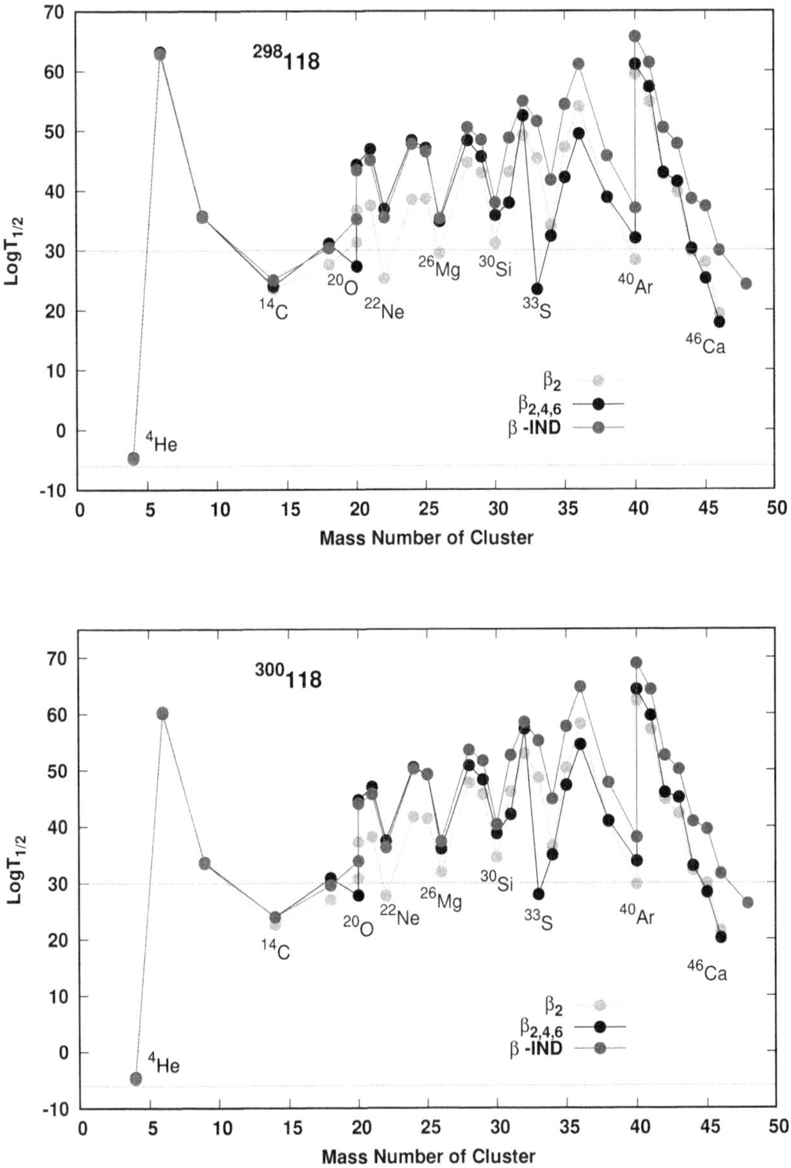

Figure 4.15: Decay half-life of different clusters for spherical case (with no deformation of parent, daughter & cluster) and with deformation (parent, daughter, cluster) for $^{298,300}118$. The dashed lines represent the experimental observation limits $10^{-6} \leq T_{1/2} \leq 10^{30}$ s.

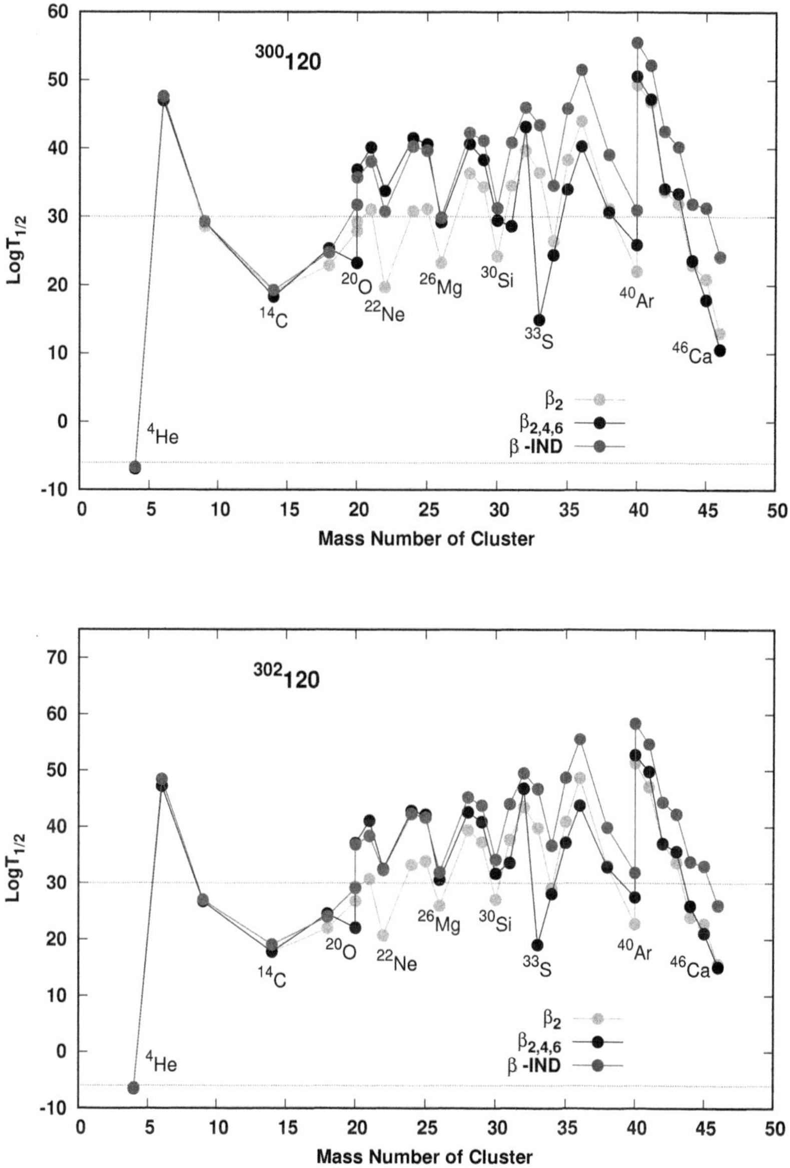

Figure 4.16: Same as in fig. 4.15 but for $^{300,302}120$.

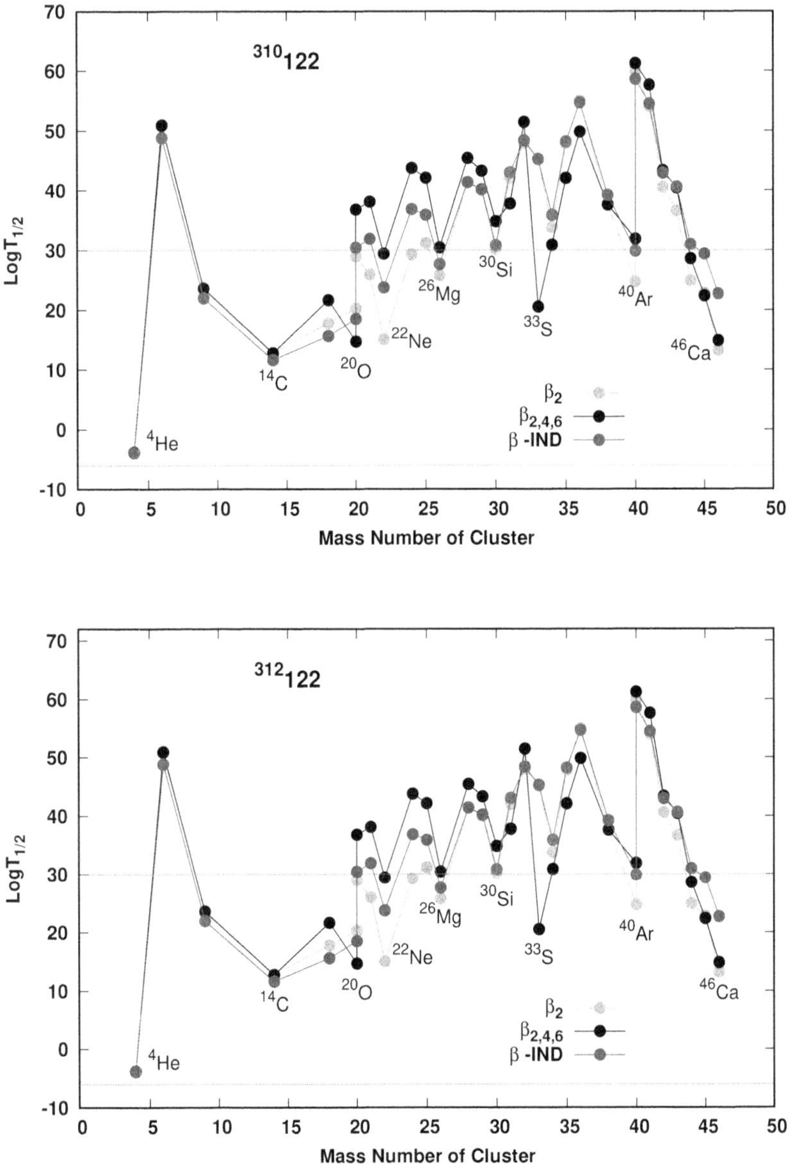

Figure 4.17: Same as in fig. 4.15 but for $^{310,312}122$.

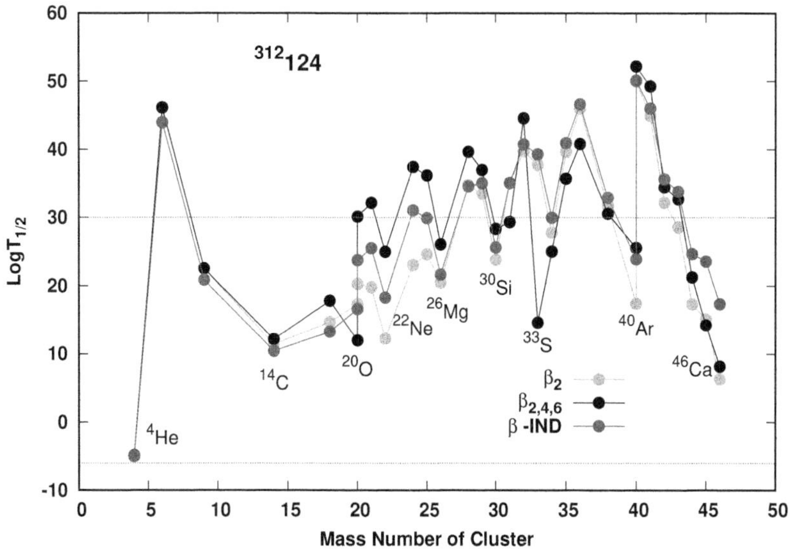

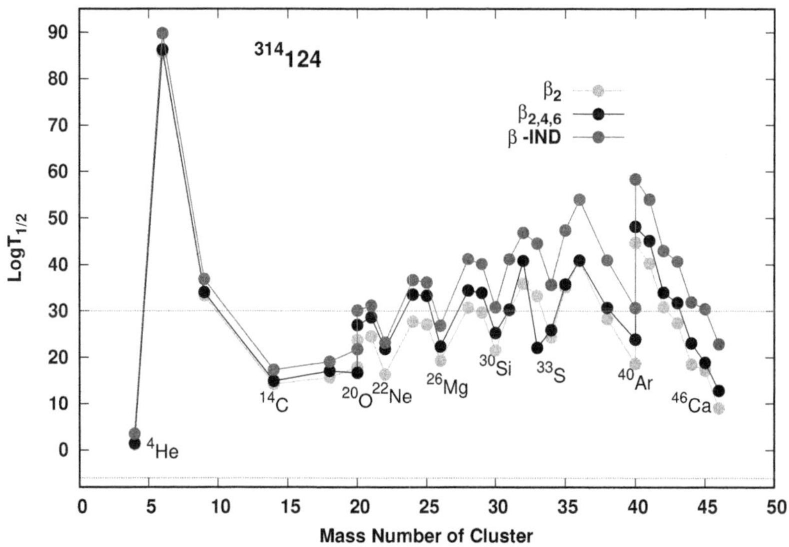

Figure 4.18: Same as in fig. 4.15 but for $^{312,314}124$.

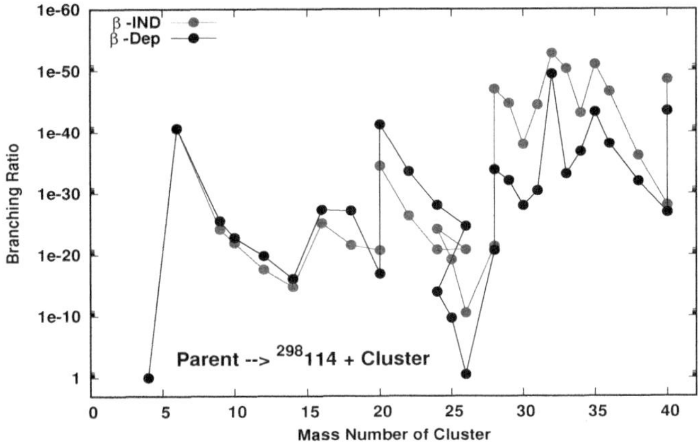

Figure 4.19: The cluster decay leading to doubly magic daughter in superheavy region. The dashed lines represent the experimental observation limits $10^{-6}s \leq T_{1/2} \leq 10^{30}s$.

4.6 Summary

- A systematic study on changes in various nuclear and interaction potential parameters by incorporating temperature in CPP model is carried out. A sharp rise in the half-life period values from N=126 to N=128 pronounces the neutron magicity at N=126.

- The cubic plus proximity model shows a good predictive power when the temperature is incorporated in the formalism.

- There is an increase in the nuclear radii and nuclear surface width when the temperature is taken into account, whereas the nuclear surface energy coefficient decreases. The increase in the height and width of the potential barrier, due to the inclusion of temperature, results in lowering the tunneling probability and higher the half-life values.

- In comparison with experimental half-life values and UDL half-life estimates, the alterations in the interaction potential with temperature enhances the accuracy of decay half-life calculated for heavy and superheavy regions. The lowest RMSE value obtained for the T-DEP CPP model is also exhibiting higher accuracy with less error. The RMSE for heavy, superheavy and both together for T-DEP CPP model comes out to be σ = 0.47, 0.72, and 0.60 respectively. Thus, this T-DEP CPP model gives better accuracy while incorporating essential parameters like temperature which may be the governing factor for the real decay mechanism.

- The inclusion of deformation modifies the interaction potential, barrier height, and width to some extent. And the turning points $V(r) = Q$ are seen to change from that of the spherical case. The cold elongated optimum configuration has a large interaction radius and smaller barrier, whereas the hot compact optimum configuration has a smaller interaction radius and larger barrier. The intermediate configurations lies in between the cold and hot configurations. Thus the deformation and orientation also govern the stability of nuclei and their decay lifetime.

- For a fixed orientation angle, the inclusion of parent deformation parameters or parent and daughter deformation parameters, reduces the barrier width and height. Cluster deformation results in a larger barrier and an increase in half-life value. With the inclusion of deformation of cluster, the calculated half-life values are in close correlation with the experimental ones for the present model.

- The non-zero quadruple and hexadecapole deformation of the nuclei involved in the decay process (parent, daughter, cluster) reduces the interaction barrier and lowers the decay half-life as compared to the spherical case and the inclusion of hexacontatetrapole deformation results in increasing the barrier and half-life values for deformed clusters. The deformation-dependent model involving parent, daughter, and cluster deformation up to hexacontatetrapole has good accuracy and predictive power.

- For cluster decay in superheavy region, the inclusion of β_2 deformation results in lowering the half-life but the trend of dips is the same as that of the spherical case as seen in the half-life versus cluster mass number curve. The inclusion of up to β_6 deformation results in new dips. This suggests that non-zero higher-order deformation parameters may favor some cluster emissions over others, like $^{20}0$ and ^{33}S, which may be a feasible observation provided their branching ratio with respect to alpha decay is good enough or with improved experimental setups.

References

Andreyev, A. N., Ackermann, D., Cagarda, P., Gerl, J., Heßberger, F., Hofmann, S., ... Wyss, R. (1999). Alpha decay of the new isotopes ^{188,189}Po. *The European Physical Journal A*, **6**, 381-385.

Bao, X. J., Zhang, H. F., Hu, B. S., Royer, G. and Li, J. Q. (2014). Systematical calculation of α-decay half-lives with a generalized liquid drop model. *Nuclear Physics A*, **921**, 85-95.

Bethe, H. (1936). An attempt to calculate the number of energy levels of a heavy nucleus. *Physical Review*, **50**, 332.

Błocki, J., Randrup, J., Świątecki, W. J. and Tsang, C. F. (1977). Proximity forces. *Annals of Physics*, **105**, 427-462.

Budaca, A. I. and Silisteanu, I. (2011). Alpha-decay as sensitive tool to derive nuclear shell structure of superheavy nuclei. *Romanian Reports in Physics*, **63**, 1147.

Daei-Ataollah, A., Ghodsi, O. N. and Mahdavi, M. (2018). Proximity potential and temperature effects on α-decay half-lives. *Physical Review C*, **97**, 054621.

Davidson, N. J., Hsiao, S. S., Markram, J., Miller, H. G. and Yiharn, T. (1994). A semi-empirical determination of the properties of nuclear matter. *Nuclear physics A*, **570**, 61–68.

Ghodsi, O. N. and Lari, F. (2014). Parametrization of barrier characteristics for deformed oriented nuclei. *Physical Review C*, **89**, 054607.

Ghodsi, O. N., Moshfegh, H. R. and Gharaei, R. (2013). Role of the saturation properties of hot nuclear matter in the proximity formalism. *Physical Review C*, **88**, 034601.

Ghorbani, F., Alavi, S. and Dehghani, V. (2020). Temperature dependence of the alpha decay half-lives of even-even Th isotopes. *Nuclear Physics A*, **1002**, 121947.

Golshanian, M., Ghodsi, O. N. and Gharaei, R. (2013). Role of surface energy coefficient and temperature of compound nucleus in the α-decay process. *Modern Physics Letters A*, **28**, 1350164.

Görgen, A., Clement, E., De France, G., Theisen, C., Wendt, A., Nyhus, H., ... others (2009). Study of oblate nuclear shapes and shape coexistence in neutron-deficient rare earth isotopes.

Gupta, R. K., Balasubramaniam, M., Kumar, R., Singh, D., Beck, C. and Greiner, W. (2005). Dynamical cluster-decay model for hot and rotating light-mass nuclear systems applied to the low-energy $^{32}S + ^{24}Mg \rightarrow ^{56}Ni^{*}$ reaction. *Physical Review C*, **71(1)**, 014601.

Gupta, R. K., Balasubramaniam, M., Kumar, R., Singh, N., Manhas, M. and Greiner, W. (2005). Optimum orientations of deformed nuclei for cold synthesis of superheavy elements and the role of higher multipole deformations. *Journal of Physics G: Nuclear and Particle Physics*, **31**, 631–644.

Hofmann, S., Ackermann, D., Antalic, S., Burkhard, H. G., Comas, V. F., Dressler, R., ... Yeremin, A. V. (2007). The reaction $^{48}Ca + ^{238}U \rightarrow ^{286}112^{*}$ studied at the GSI-SHIP. *The European Physical Journal A*, **32**, 251-260.

Hosseini, S., Hassanabadi, H. and Zarrinkamar, S. (2018). A comparative analysis of alpha-decay half-lives for even–even ^{178}Pb to ^{234}U isotopes. *Nuclear Physics A*, **970**, 259 - 271.

Hosseini, S. S., Hassanabadi, H. and Zarrinkamar, S. (2016). Optimal temperature for alpha-decay half-lives with yukawa proximity potential. *International Journal of Modern Physics E*, **25**, 1650109.

Ismail, M. and Adel, A. (2011). Orientation dependent behavior of the coulomb barrier parameters for deformed–deformed nuclei. *Nuclear Physics A*, **859**, 1-15.

Ismail, M. and Seif, W. (2011). Prediction of accidental cancellation of different deformation components and optimum fusion orientations. *Nuclear Physics A*, **872**, 25-41.

Jaqaman, H. R. (1989). Instability of hot nuclei. *Physical Review C*, **40**, 1677–1684.

Kaur, G., Sandhu, K. and Sharma, M. K. (2018). Effect of deformations on the compactness of odd-Z superheavy nuclei formed in cold and hot fusion reactions. *Nuclear Physics A*, **971**, 95-112.

Keli, A., Natowitz, J. B. and Schmidt, K. H. (2006). *Dynamics and thermodynamics with nuclear degrees of freedom.* Springer Berlin Heidelberg.

Kruppa, A. T., Bender, M., Nazarewicz, W., Reinhard, P.-G., Vertse, T. and Ćwiok, S. (2000). Shell corrections of superheavy nuclei in self-consistent calculations. *Physical Review C*, **61**, 034313.

Le Couteur, K. J. and Lang, D. W. (1959). Neutron evaporation and level densities in excited nuclei. *Nuclear Physics A*, **13**, 32 - 52.

Malhotra, N. and Gupta, R. K. (1985). Proximity potential for deformed, oriented collisions and its application to $^{238}U + ^{238}U$. *Physical Review C*, **31**, 1179–1183.

Moller, P., Sierk, A., Ichikawa, T. and Sagawa, H. (2016). Nuclear ground-state masses and deformations: Frdm(2012). *Atomic Data and Nuclear Data Tables*, **109-110**, 1.

Morinaga, H. (1956). Interpretation of some of the excited states of 4*n* self-conjugate nuclei. *Physical Review*, **101**, 254–258.

Morita, K., Morimoto, K., Kaji, D., Akiyama, T., Goto, S., Haba, H., … Yoshida, A. (2007). Observation of second decay chain from ^{278}113. *Journal of the Physical Society of Japan*, **76**, 045001.

Myers, W. D. and Swiatecki, W. J. (1966). Nuclear masses and deformations. *Nuclear Physics*, **81**, 1-60.

Naveya, G., Santhosh Kumar, S., Philominraj, S. I. A. and Stephen, A. (2019). Study on particle and cluster decay of superheavy nuclei Z = 130 - 144 using cubic plus proximity potential with improved transfer matrix method. *International Journal of Modern Physics E*, **28**, 1950051.

Oganessian, Y. T., Utyonkov, V., Lobanov, Y., Abdullin, F., Polyakov, A., Shirokovsky, I., … Lougheed, R. (2004a). Heavy element research at dubna. *Nuclear Physics A*, **734**, 109-123.

Oganessian, Y. T. and Utyonkov, V. K. (2015). Super-heavy element research. *Reports on Progress in Physics*, **78**, 036301.

Oganessian, Y. T., Utyonkov, V. K., Dmitriev, S. N., Lobanov, Y. V., Itkis, M. G., Polyakov, A. N., … Eichler, R. (2005). Synthesis of elements 115 and 113 in the reaction $^{243}Am + ^{48}Ca$. *Physical Review C*, **72**, 034611.

Oganessian, Y. T., Utyonkov, V. K., Lobanov, Y. V., Abdullin, F. S., Polyakov, A. N., Sagaidak, R. N., … Lougheed, R. W. (2006). Synthesis of the isotopes of elements 118 and 116 in the ^{249}Cf and $^{245}Cm + ^{48}Ca$ fusion reactions. *Physical Review C*, **74**, 044602.

Oganessian, Y. T., Utyonkov, V. K., Lobanov, Y. V., Abdullin, F. S., Polyakov, A. N., Sagaidak, R. N., … Wilk, P. A. (2007). Synthesis of the isotope $^{282}113$ in the $^{237}Np + ^{48}Ca$ fusion reaction. *Physical Review C*, **76**, 011601.

Oganessian, Y. T., Utyonkov, V. K., Lobanov, Y. V., Abdullin, F. S., Polyakov, A. N., Shirokovsky, I. V., … Lougheed, R. W. (2004b). Measurements of cross sections for the fusion-evaporation reactions $^{244}Pu(^{48}Ca,xn)^{292-x}114$ and $^{245}Cm(^{48}Ca,xn)^{293-x}116$. *Physical Review C*, **69**, 054607.

Oganessian, Y. T., Utyonkov, V. K., Lobanov, Y. V., Abdullin, F. S., Polyakov, A. N., Shirokovsky, I. V., … Vesnovskii, S. P. (2005). Measurements of cross sections and decay properties of the isotopes of elements 112, 114, and 116 produced in the fusion reactions $^{233,238}U$, ^{242}Pu, and $^{248}Cm + ^{48}Ca$ [Phys. Rev. C 70, 064609 (2004)]. *Physical Review C*, **71**, 029902.

Oganessian, Y. T., Utyonkoy, V. K., Lobanov, Y. V., Abdullin, F. S., Polyakov, A. N., Shirokovsky, I. V., … Lougheed, R. W. (2004). Experiments on the synthesis of element 115 in the reaction $^{243}Am(^{48}Ca,xn)^{291-x}115$. *Physical Review C*, **69**, 021601.

Oganessian, Y. T., Yeremin, A. V., Popeko, A. G., Bogomolov, S. L., Buklanov, G. V., Chelnokov, M. L., … Morita, K. (1999). Synthesis of nuclei of the superheavy element 114 in reactions induced by ^{48}Ca. *Nature*, **400**, 242-245.

Pahlavani, M. and Rahimi Shamami, S. (2020). Clusters decay half-live of various heavy deformed nuclei with mass numbers in the range $221 \leq A \leq 242$. *Chinese Journal of Physics*, **66**, 733-744.

Pahlavani, M. R., Alavi, S. A. and Tahanipour, N. (2013). Effect of Nuclear Deformation on the Potential Barrier and Alpha-Decay Half-Lives of Superheavy Nuclei. *Modern Physics Letters A*, **28**, 1350065.

Pahlvani, M. R., Alavi, S. A. and Tahanipour, N. (2013). Effect of nuclear deformation on the potential barrier and alpha-decay half-lives of superheavy nuclei. *Modern Physics Letters A*, **28**, 1350065.

Pei, J. C., Xu, F. R., Lin, Z. J. and Zhao, E. G. (2007). α-decay calculations of heavy and superheavy nuclei using effective mean-field potentials. *Physical Review C*, **76**, 044326.

Preetha, P. and Santhosh Kumar, S. (2020). Nuclear Level Density and the Structural Dynamics of Rotating Superheavy Nucleus Z = 117. *Brazilian Journal of Physics*, **50**, 346-362.

Royer, G. and Mignen, J. (1992). Binary and ternary fission of hot and rotating nuclei. *Journal of Physics G: Nuclear and Particle Physics*, **18**, 1781.

Salehi, M. and Ghodsi, O. N. (2013). The influence of the dependence of surface energy coefficient to temperature in the proximity model. *Chinese Physics Letters*, **30**, 042502.

Santhosh, K., Biju, R. and Sahadevan, S. (2009). Stability of 244-260 Fm isotopes against alpha and cluster radioactivity. *Pramana*, **73(6)**, 1059–1072.

Santhosh, K. and Joseph, A. (2002a). Effect of parent and daughter deformation on half-life time in exotic decay. *Pramana – Journal of Physics*, **59**, 679-684.

Santhosh, K. and Joseph, A. (2002b). Effect of parent and daughter deformation on half-life time in exotic decay. *Pramana*, **59**, 679-684.

Santhosh, K., Subha, P. and Priyanka, B. (2016). Cluster decay of 112-122Ba isotopes from ground state and as an excited compound system. *Pramana – Journal of Physics*, **86**, 819–836.

Soylu, A. and Evlice, S. (2015). Deformation effects on cluster decays of radium isotopes. *Nuclear Physics A*, **936**, 59-73.

Soylu, A. and Qi, C. (2021). Extended universal decay law formula for the α and cluster decays. *Nuclear Physics A*, **1013**, 122221.

Umai Parvathiy, J. (2016). Properties of super heavy elements in trans actinide region. *Ph.D. Thesis*.

Weisskopf, V. (1937). Statistics and nuclear reactions. *Physical Review*, **52**, 295.

Wong, C. Y. (1973). Interaction barrier in charged-particle nuclear reactions. *Physical Review Letters*, **31**, 766–769.

Zanganah, V., Akrawy, D. T., Hassanabadi, H., Hosseini, S. and Thakur, S. (2020). Calculation of α-decay and cluster half-lives for $^{197-226}$Fr using temperature-dependent proximity potential model. *Nuclear Physics A*, **997**, 121714.

Zanganeh, V. and Wang, N. (2014). Temperature-dependent potential in alpha-decay process. *Nuclear Physics A*, **929**, 94 - 101.

CHAPTER 5

SCREENED ALPHA DECAY IN TERRESTRIAL AND ASTROPHYSICAL CONDITIONS

5.1 Introduction

During the big-bang epoch, the primordial elements, hydrogen and helium appeared first and subsequently such elements formation occurred in various other cosmic events. The huge gas clouds of hydrogen and helium called nebulae falling under gravity give birth to stars, which settle into an equilibrium structure when gravity is balanced by pressure from nuclear processes and gases. Stars almost spend a significant portion of their lives in hydrostatic equilibrium.

In the stellar interior heavy element formation occurs via nuclear fusion. Elements up to carbon are created in low mass stars ($< 8M_{\odot}$), which at the end of their life on running out of hydrogen to fuse will become white dwarfs. In the interior of massive stars ($> 8M_{\odot}$), even carbon gets fused further and heavier elements such as iron, nickel, and cobalt are formed. Elements heavier than iron are formed in the phenomenon where neutrons excess nuclei and the free neutrons available combine together to form a nucleus, which is in general radioactive in nature and further decays to form stable heavy isotopes. Processes such as the s-process which is considered to occur in the last evolutionary stages of stars, and the r-process in supernova explosions and binary neutron star collisions, where the substantial neutron flux is available lead to heavy element formation in astrophysical events beyond Fe.

Recent research from the observations of the chemical composition of stars reported in Reticulum II (Ret II), a satellite of the Milky way, in 2016 (Frebel and Beers, 2018) suggested the likelihood of neutron-star merger events being the main source of heavy element production instead of the supernova explosion. Heavy element formation signatures were observed in binary merger GW170817, detected at first in gravitational wave regime by the Laser-Interferometer Gravitational-wave Observatory (LIGO) and later its electromagnetic counterpart was also seen by space and ground-based telescope (Abbott et al., 2017). Also, the upcoming Facility for

Rare Isotope Beam (FRIB) have its contributions pointing towards discovering the properties of rare isotopes, understanding the origin of nuclear matter, the nuclear astrophysics, mapping the nuclear landscapes, the fundamental forces that bind the nuclei, nuclear medicine, and nuclear waste management (Balantekin et al., 2014; Hausmann et al., 2013).

Different areas of research concerning, astronomy, astrophysics, theoretical and experimental nuclear physics are making progress to understand the r-process phenomenon which is responsible for element formation in the observable universe. The theoretical and simulation work on neutron star and supernova collapse together with multi-messenger surveys of electromagnetic signals and neutrinos can help in obtaining crucial information about physics of supernova and merger events. In this regard, nuclear physicists are in the quest to obtain accurate predictions and estimates for masses, nuclear interaction cross-sections, decay rates of heavy and superheavy nuclei, and also finding the stable and unstable nuclei in these mass regions. In this regard, a study of decay properties of nuclei at different environments is essential.

At this juncture, it is important to discuss about the electron screening. As we know, nuclear waste contains highly radioactive materials of a long lifetime. To deal with disposable managements of transuranium nuclear waste materials embedding them in metals at the low temperature indicated an enhanced electron screening during the decay process of elements and a shift in decay constant is noticed due to this screening, thereby lowering its life-time (Zinner, 2007). Such an electron screening can also be implemented and studied on nuclear fusion reaction rates occurring in stellar processes (Czerski et al., 2001; Potekhin and Chabrier, 2013). In an astrophysical environment, strong magnetic fields play a crucial role in most of the processes and are likely to influence the alpha decay process of heavy elements in the neighborhood of magnetars. Also, the dense plasma matter present in the stellar interior, interstellar medium, and neighborhood of neutron stars/black holes can induce significant changes in nuclear decay properties. This chapter focuses on studying the decay rate of heavy and superheavy nuclei in different external environments, the terrestrial and astrophysical one. In the previous chapters, the nuclear temperature and deformation are taken as intrinsic properties of nuclei and are incorporated in the study of decay properties which modifies the nuclear radii and interaction potential, and thus changing the decay lifetime of nuclei. In the present chapter, sec 5.2 deals about the decay processes occurring in the atomic environments, sec. 5.3 & 5.4 investigates the effect of strong magnetic fields and dense plasma in stellar environments followed by its results and detailed discussions.

5.2 Decay in atomic environment

In bare nucleus, a decay process is simply a daughter and fragment formation from the parent nucleus. But the decay process involving atoms should always be associated with the electrons of parent, daughter and fragment particles. Here, the electrons are not just in silent background to the decay process but the parent electrons participate in the decay process where they re-organize themselves from parent to daughter and alpha/cluster atoms. Considering such atomic nature to the participating elements, the Q-value of the decay process will be of the following form,

$$Q = Q_{nuc} + BE_e(Z) - BE_e(Z-2) - BE_e(2), \tag{5.103}$$

where $BE_e(Z)$, $BE_e(Z-2)$ and $BE_e(2)$ are total electron binding energies of the participating atoms taken into consideration. The energy released solely as a result of the nuclear mass difference between the initial and final decay products is Q_{nuc}. A simple relation to Q-value accounting for the electron screening in decay process is, $Q = \Delta M_p - (\Delta M_d + \Delta M_\alpha) + k(Z_p^\varepsilon - Z_d^\varepsilon)$, where the term $k(Z_p^\varepsilon - Z_d^\varepsilon)$ is the correction and denotes the total binding energy of the Z electrons present in the atom. The electrons surrounding the nucleus are bound in the atoms hence the binding energy gets associated with them. The value of parameters are $k = 8.7$ eV and $\varepsilon = 2.517$ for nuclei having $Z \geq 60$, and $k = 13.6$ eV and $\varepsilon = 2.408$ for nuclei having $Z < 60$ (Santhosh et al., 2016), this values are extracted from the neutral atom electron binding energy studies of Ref. (Huang et al., 1976).

Apart from altering the Q-value of the decay process, the electrons involved in the decay process can suppress the Coulomb interaction between the daughter and emitted fragment by changing the interaction barrier. The dynamics which occur in the decay process of the neutral parent atom (^A_ZX) in astrophysical reactions where the electron rearrange themselves around the neutral daughter and neutral alpha particle is likely to be quite complex. However, on the basis of two possible approaches, this can be described as mentioned in ref. (Liolios, 2003). Such as, (i) In the adiabatic limit, the electron clouds are assumed to be rapidly re-arranging themselves during the decay process resulting in the parent, daughter, and alpha particles being neutral at all times. (ii) In the sudden limit, during the decay process, the electron cloud of the parent is considered to be unchanged such that the daughter nucleus is screened by the same electron cloud of parent nuclei and the alpha particle emitted is fully ionized. In both these limits calculations using Thomas-Fermi (TF) approach, a quantum mechanical theory for structures involving electron clouds of many-body system, yield a simple relation which connects the screened and non-screened decay process half-lives, which can be written as,

$$T_{1/2}^{SC}(Z,A,Q_{nuc}) = \frac{T_{1/2}^{NSC}(Z,A,Q_{nuc})}{f_\alpha(Z,A,Q_{nuc})}. \tag{5.104}$$

The factor f_α is called as screening enhancement factor (SEF). The screening factors description for both the adiabatic and sudden limit is seen to be nearly same behavior as in ref. (Liolios, 2003). For ease of calculation, the simple relation provided by the sudden limit $f_\alpha(E)$ is considered, and is given by,

$$f_\alpha(E) \simeq \exp\left[-\frac{3.856(Z-2)^{7/3}}{Q_{nuc}^{3/2}}\left(\frac{A}{A-4}\right)\left[S(q)+\frac{q}{x_0(q)}\right]\right], \tag{5.105}$$

where

$$q = 1 - \frac{\text{no. of electrons}}{\text{no. of protons}} \tag{5.106}$$

is the degree of ionization. The values of $S(q), x_0(q)$ are defined in (Liolios, 2001). For neutral atoms where the charges of electrons and protons are in balance, the value of $q = 0$ and $S(q) = -1.588$, then the Thomas- Fermi enhancement factor will be,

$$f_{TF}(E) \simeq \exp\left[\frac{6.1233(Z-2)^{7/3}}{Q_{nuc}^{3/2}}\left(\frac{A}{A-4}\right)\right]. \tag{5.107}$$

The above relation is simple to implement and is employed for finding the screening enhancement factor. The screened half-life obtained when the half-lives determined under non-screened conditions in earlier chapters is divided with the enhancement factor, will be helpful in studying the screening effect in astrophysical nuclear reactions.

5.3 Strong magnetic fields

Neutron stars are compact objects formed when progenitor stars with masses of $> 8\,M_\odot$ explode in supernovae. The neutron stars are rapidly rotating ones and are extremely dense ($\sim 10^{15}$ g/cm^3) and possess strong magnetic field at their surface ($10^8 - 10^{13}$ G) (Glendenning, 1997). These stars were first theoretically predicted by Baade and Zwicky (Baade and Zwicky, 1934), and their observational evidence includes radio pulsars, accretion of pulsars in low mass X-ray binaries, and neutron stars in binary merger where detected in LIGO in gravitational waves. A subgroup of neutron stars having magnetic fields larger than typical ones is called Magnetars (10^{14-16} G) (Potekhin, 2010). During the birth of neutron star in supernova explosion of massive stars, the neutron flux present results in heavy element formation. Also, accretion of matter from its companion in binary results in the matter and angular momentum deposition in neutron stars (Beard and Wiescher, 2003; Rehm, 2016). During this episode the heavy elements may find themselves in the neighborhood of neutron stars or magnetar's magnetic fields. The influence of such a strong magnetic fields is ascribed for alpha decaying elements. When heavy neutral atom decay, the atomic electron clouds are compressed by the external magnetic field resulting in screened Coulomb potential, which can be described by (Liolios, 2001),

$$\Phi_{sc}(r) = \frac{Ze}{r}\phi\left(\frac{r}{R_B}\right), \tag{5.108}$$

where $R_B = 55133\ Z^{1/5}b^{-2/5}$ fm is the scaling parameter, $b = B/B_0$ with $B_0 = 2.351 \times 10^9$ G. The universal function $\phi(x)$ is the solution obtained from Kadomstsev's standard differential equation (Kadomtsev, 1970; Banerjee et al., 1974),

$$\frac{d^2\phi(x)}{dx^2} = (x\phi)^{1/2}, \phi(0) = 1, \phi'(0) = -0.938965, \tag{5.109}$$

with $x = r/R_B$. The strong magnetic screened Coulomb potential using Thomas-Fermi sudden limit will take the form,

$$V^{SC}(r,B) = \frac{2(Z-2)e^2}{r}\phi\left(\frac{r}{R_B}\right), \tag{5.110}$$

where $R_B = 55133\,(Z-2)^{1/5}\,b^{-2/5}$ fm. The screening function can be obtained by solving numerically or analytically. The analytical solution to the screening function is given by (Wan et al., 2015),

$$\phi(x) = 1 + Sx + \frac{4}{5}x^{2.5} + \frac{2}{35}Sx^{3.5} - \frac{1}{126}S^2x^{4.5}, \tag{5.111}$$

where $S = \phi'(0) = -0.938965$. This model is valid for neutral atoms under the condition that $Z^{4/3} \ll b \ll 2Z^3$. In fig. 5.1 the solution of Kadomstev's equation from $r = 0$ to the outer turning point is presented. It is evident that it is a decreasing function, the touching distance is at $r \approx 10$ fm and from which the Coulomb potential acts between a daughter and an alpha particle. At $r = 0$, $\phi(r/R_B)$ has value 1, and at touching distance ($r_t \approx 10$ fm) it is about 0.99 for strong magnetic fields ($10^{13} - 10^{15}$ G) and is about 0.96 for very strong magnetic field 10^{16} G. The factor by which the magnetic field reduces Coulomb potential as compared to the non-screened case can be inferred from the value of $\phi(r/R_B)$. The screened Coulomb potential reduces by a factor of 0.96 at a touching distance and by 0.86 at a turning point for a 10^{16} G environment, whereas it gets reduced by 0.99 at a touching distance and 0.98 at the turning point for 10^{13} G environment.

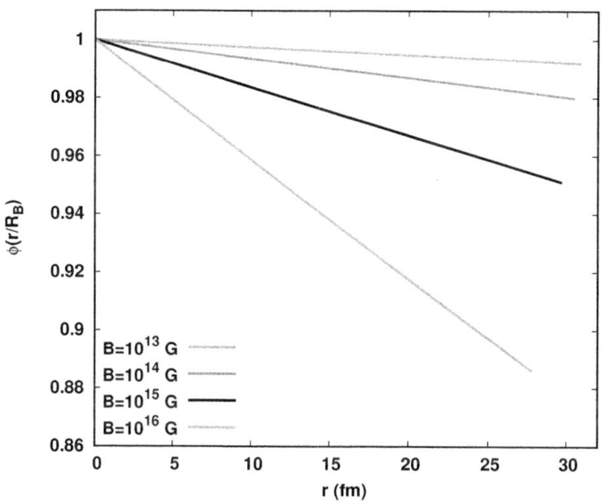

Figure 5.1: The universal function $\phi\left(\frac{r}{R_B}\right)$ which is a correction factor to Coulomb potential, obtained from solution of Kadomstsev's equation for different values of magnetic field for the decay process $^{294}118 \rightarrow ^{290}116 + \alpha$.

5.4 Dense astrophysical plasma

Plasma are superheated matter consisting of positive and negative charged particles where charge neutrality is maintained and is produced when atoms get ionized in the gas. The astrophysical plasma is found in the stellar interior, around the neutron star and black holes in the accretion disk and/or jets, and also in the interstellar medium. The participating nuclei in thermonuclear reactions polarize the plasma present around it, which results in modification of the Coulomb potential. This modified potential is the Yukawa type, which is comparatively smaller than that of free-space (or vacuum) causing easy tunneling and hence reducing the decay lifetime. The effect of screening on nuclear reactions occurring in sun and stars, in the big bang nucleosynthesis and supernova collapse, has been widely studied (Arafune and Fukugita, 1992; Mussack and Däppen, 2011; Potekhin and Chabrier, 2013; Wang et al., 2011; Liu et al., 2009). The first work on screening factors obtained by Salpeter is good at low densities (Salpeter, 1954), but for higher densities Mitler's approach is well suited (Mitler, 1977). Mitler obtained screened Coulomb potential in the context of fusion reactions, with suitable modification to the approach and which can be extended to study the decay processes (Liolios, 2003). The screened Coulomb potential for daughter and alpha particle in the interior of plasma is given by

$$V_c^{sc}(r) = \frac{2(Z-2)}{r} - C_o + C_1 r^2, \qquad r < r_0 \tag{5.112}$$

$$V_c^{sc}(r) = \frac{2(Z-2)C}{r}, \qquad r > r_0, \tag{5.113}$$

with $x = \frac{r_0}{R_D} = \left(\frac{3(Z-2)}{4\pi n_e R_D^3} + 1\right)^{1/3} - 1$ where R_D is the Debye-Huckel radius, n_e is the average electron number density in the plasma and the coefficients are $C_0 = 2\pi e n_e R_D^2 x(x+2)$, $C_1 = \frac{2}{3}\pi e n_e$, and $C = 4\pi n_e r_0 R_D^2 e^{r_0/r_D}$. The Debye-Huckel radius can be obainted using (Somov, 2012),

$$R_D = \sqrt{\frac{k_B T}{4\pi e^2 n_e}}, \tag{5.114}$$

which depends on the electron number density n_e and temperature T of the matter. The Debye-Huckel radius denotes the scale/distance over which the electrostatic screening in plasma and conductors persists.

5.5 Results and discussion

Incorporating the strong magnetic field and dense plasma screening to the interaction potential, the post-scission region potential of cubic plus proximity model takes the following form,

$$V_{ext}(r,\theta,\phi) = V_n(r,\theta,\phi) + V_C^{sc}(r,\theta,\phi) + V_l(r,\theta,\phi). \tag{5.115}$$

Depending on the touching distance of post-scission potential $V_{ext}(r_t,\theta,\phi)$, the pre-scission potential gets altered. The change in interaction potential due to the magnetic field and dense plasma environment is shown in fig. 5.2. It is evident from the figure that the addition of a magnetic field (10^{15} G) results in a lowering of the interaction potential, and the barrier height and width also get reduced. Similar such effect is seen for decay in the dense plasma environment (10^9 g/cm^3 at 10^8 K) (Meyer, 1989). Also, a shift in the turning point is seen under the influence of different environments, enhancing the decay rate. Studies with 116 isotopes of heavy and superheavy nuclei in the range $Z = 84 - 118$ are carried out using experimental Q-values and decay lifetimes of bare nuclei obtained from chapter 4, wherein for a neutral atom under the effect of atomic environment, magnetic field (10^{14-16} G) and dense plasma matter with a density as that of neutron star surface ($\rho = 2\times10^{9-10}$ g/cm^3 at temperature $T = 3\times10^8$ K) (Potekhin, 2010; Tsuruta, 2016; Meyer, 1989) , and its screened half-lives are deduced.

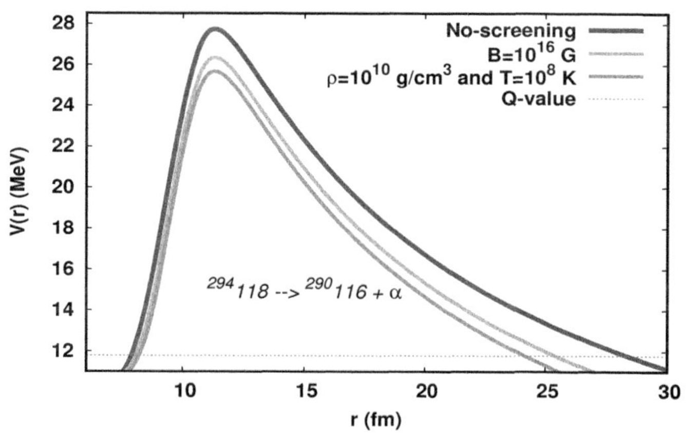

Figure 5.2: The non-screened and screened potential is shown for decay process $^{294}118 \rightarrow^{290} 116+\alpha$. The screened potential is for magnetic field of 10^{16} Gauss.

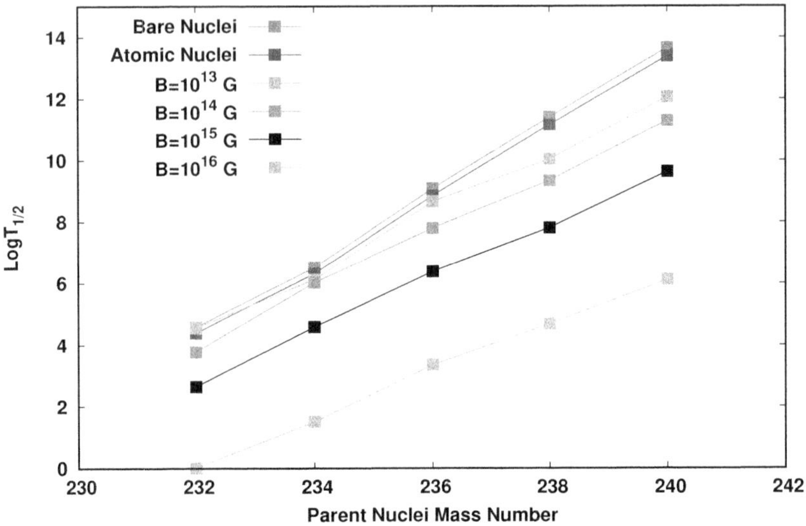

(a) **Z=94**

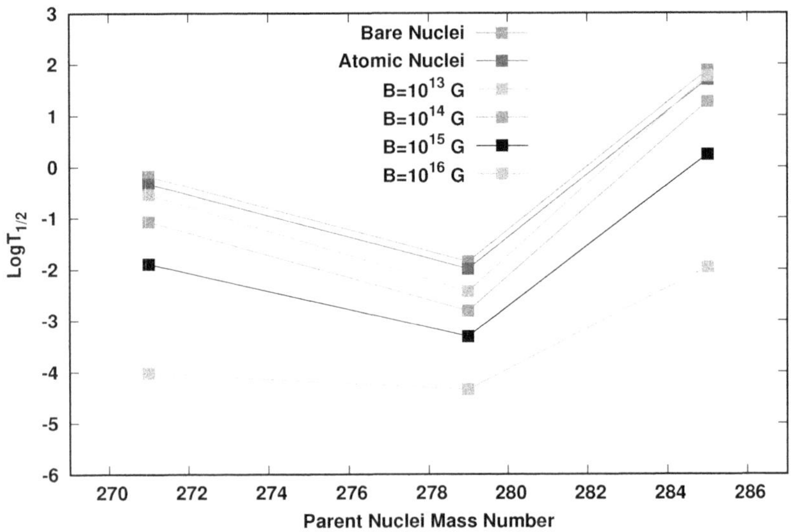

(b) **Z=114**

Figure 5.3: Alpha decay half-lives of (a) Z=94 and (b) Z=114 isotopes calculated in non-screened, atomic and strong magnetic field environments.

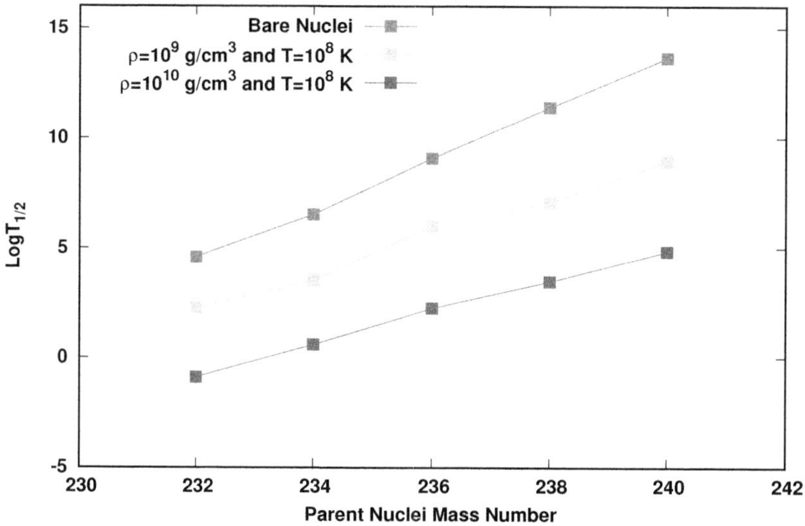

(a) **Z=94**

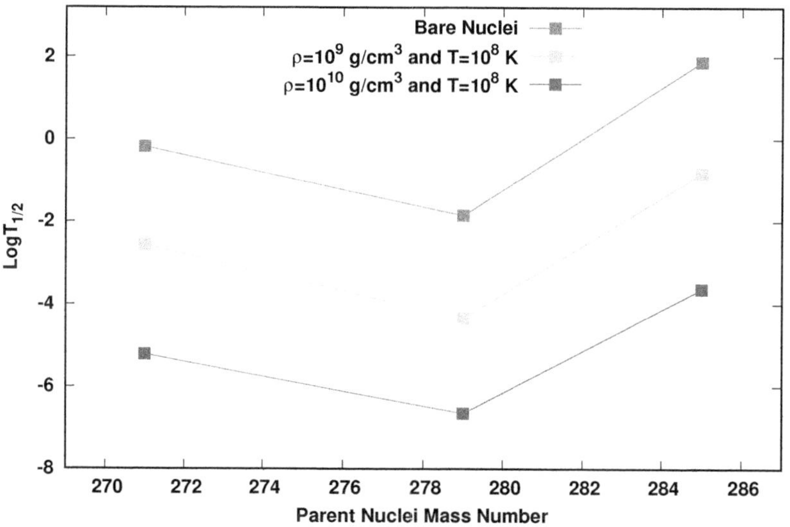

(b) **Z=114**

Figure 5.4: Same as in fig. 5.3 but for bare nuclei and in screened dense plasma environment.

The half-life versus parent nuclei mass number is shown for alpha decay for the isotopes of Pu (Z=94) and Fl (Z=114) in fig. 5.3. The behaviour of alpha decay half-lives in the atomic environment and bare nuclei seems to be close to each other, which signifies that electrons in the atomic environment gets screened and enhances the half-life by a small proportion. Also, in the strong magnetic fields, where stronger the magnetic field greater is the change in half-life, an increase in decay life-time by a few orders with B= 10^{16} G fig. 5.3 and 5 orders change is seen from that of bare nuclei, is substantial. A comparison on the half-life of bare nuclei and plasma environment for the isotopes of Pu and Fl is presented in fig. 5.4. The dense plasma reduces the half-life by a substantial amount, upto 8-10 orders is seen for a density of $\approx 10^{10}$ g/cm^3 at temperature $\approx 10^8$ K, whereas density of 10^9 g/cm^3 at the same temperature will reduce the half-life upto 5 orders. A dip near to $A = 278$ for Z=114 (fig. 5.3 (b)) indicates rapid decay by parent nuclei to attain magic number configuration $^{278}_{114}Fl_{164}$. Also from fig. 5.4(b) the signature of magicity at $N = 164$ associated with alpha decay process in the plasma environment doesn't seem to alter.

The screening enhancement factor for the atomic, strong magnetic field, and dense plasma environments as a function of parent nuclei atomic number is shown in fig. 5.5 - 5.7, the vertical axis scale indicates the order up to which f_α ranges in these environments for different nuclei. The screening factor is greater than 1 for all nuclei, which suggests the decay half-life gets reduced in a different environment ie., the decay rate gets enhanced. In the atomic environment for ^{232}Th (Z=90) as seen in fig. 5.5, the maximum value of the screening factor lies between 1-2.5. For the strong magnetic field environment (Potekhin, 2010) with B= 10^{14} G (fig. 5.5) and 10^{15} G (fig. 5.6), the screening enhancement factor in the range of $1 - 10^3$ and $1 - 10^4$ has a maximum screening value for ^{240}Pu (Z=94) among heavy isotopes. When the value of the magnetic field increases to B= 10^{16} G (fig. 5.6) with a screening factor in the range $1 - 10^9$, the maximum values are seen for ^{232}Th and ^{240}Pu. The screening enhancement in dense plasma (fig. 5.7) suggests that for the same temperature of matter with an increase in density, the screening factor changes significantly. The screening factor for dense plasma matter having temperature $T = 3 \times 10^8$ K and matter density 2×10^9 g/cm^3 in the range $1 - 10^7$ and for matter density 2×10^{10} g/cm^3 in $1 - 10^{10}$ range, ^{232}Th is said to possess a high screening value. It is to be noted that, the screening enhancement for super-heavy isotopes ($Z > 104$) is lower than that of heavy isotopes for the studied environments, and the maximum enhancement factor is observed

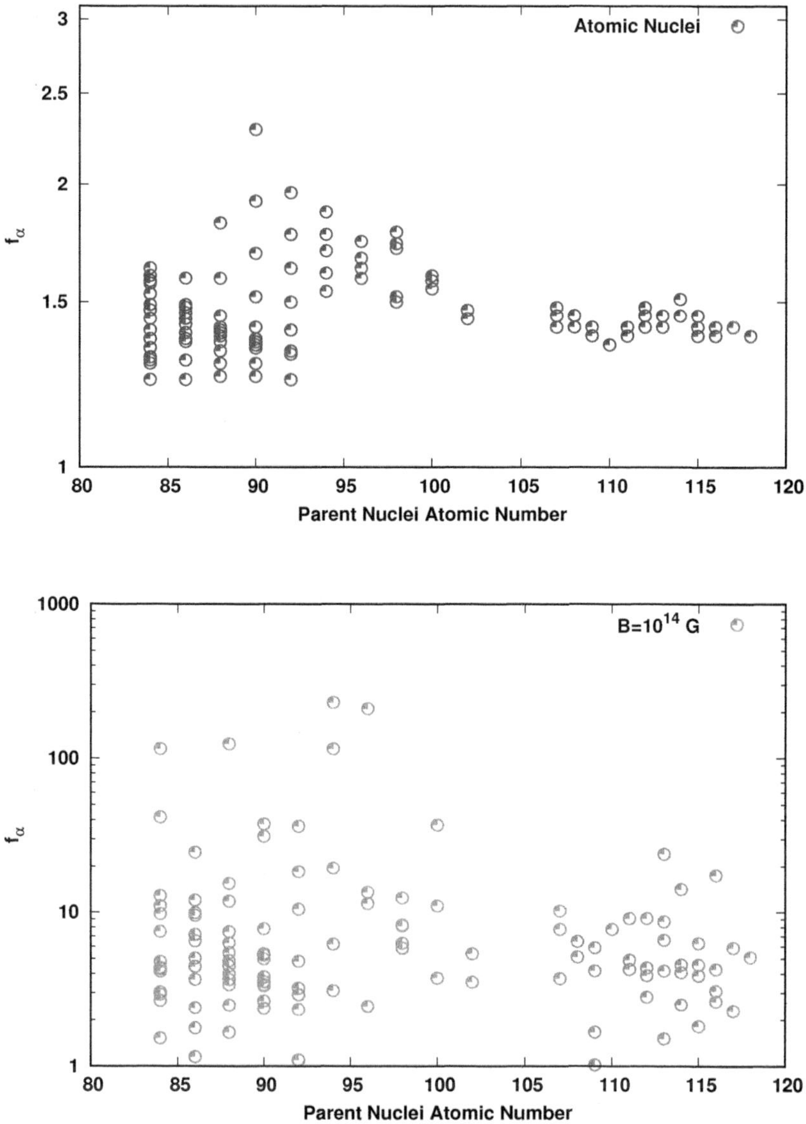

Figure 5.5: Screening enhancement factor for heavy and superheavy isotopes (Z= 84-118) undergoing alpha- decay process in atomic, strong magnetic field $B = 10^{14}$ G environment.

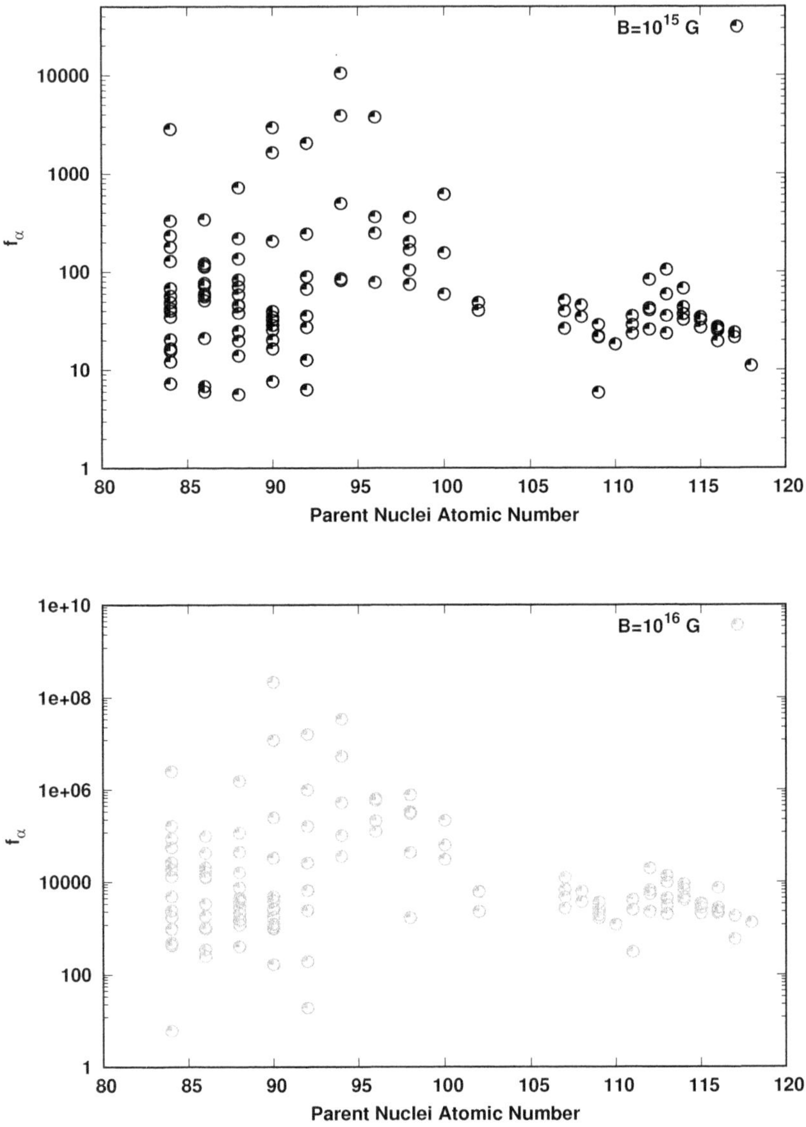

Figure 5.6: Same as in fig. 5.5 in strong magnetic field but for $B = 10^{15}$ G and $B = 10^{16}$ G environment.

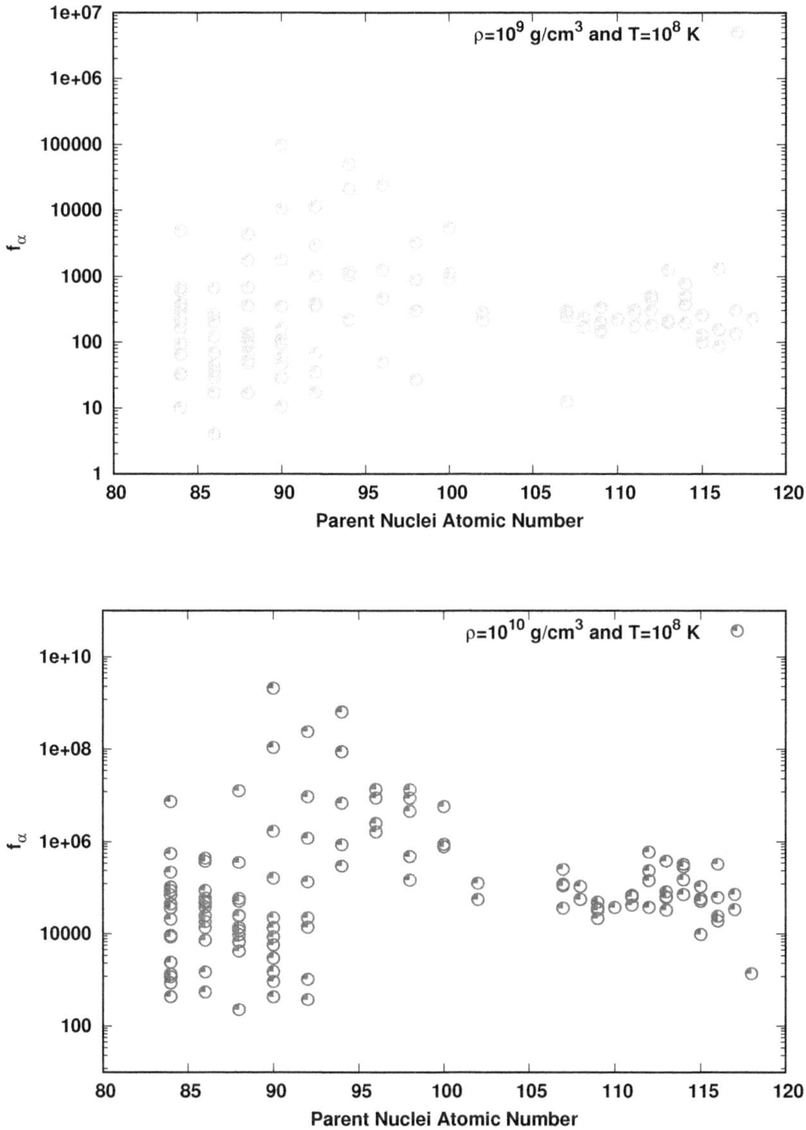

Figure 5.7: Same as in fig. 5.5 in astrophysical dense plasma environment.

for the isotopes of $^{284}113$, $^{282}112$ and $^{291}116$.

The astrophysical events rich in neutron flux are the sites where in the first place formation of neutron-rich nuclei occurs (by r-process). These nuclei are located far from the beta stability line and close to the neutron drip line. To attain stable configuration, these elements further undergo primarily beta decay to attain nuclear configuration existing in the beta stability region. Further decay of these elements made them to reach comparatively stable isotopes. The r-process simulation studies on chemical abundance usually overlook the alpha decay and consider only beta decay and photon disintegration. Recent studies incorporating alpha decay have suggested that this decay mode can play a crucial role in the abundance data where the peak represents the product elements formed post alpha decay (Schilbach et al., 2018). The alpha decay when subjected to astrophysical environments, is an active participant in the in the element formation and destruction together with other modes due to the reduction of lifetime (Liolios, 2001). In the present study, incorporating these screening potentials/factors under different terrestrial and astrophysical environments into the Cubic Plus Proximity model makes this model the very first model of its kind to study the pre-scission and post-scission regions for examining the screened decay lifetimes, which is necessary for completeness and accuracy. The calculations bring out the magnitude of the screening enhancement factor under different environmental conditions and also indicate that the screening in heavy isotopes is greater in comparison to the superheavy region and that these conditions influence different elements in a different manner leading to various levels of enhanced decay rates. The results obtained may have a direct consequence on the role played by alpha decay instability and decay properties of heavy/super heavy nuclei under various environmental conditions and also towards the study of chemical abundance existing in these astrophysical event sites, ranging from the late stages of stars, our solar system which has undergone different stages of recycling and evolution, and the newly born and old galaxies within our universe.

5.6 Summary

- The heavy and superheavy elements far from the beta stability line and those close to the beta stability line are formed in astrophysical events where substantial neutron flux is present, like the late stages of stars, supernova explosions, and binary neutron star mergers. These isotopes decay to stable ones create bright electromagnetic radiations called nova or kilonova.

- The elements formed in such astrophysical events are in close proximity to neutron stars which possess strong magnetic fields and densities. Also, in binary star systems, the mass lost by a companion star is accreted by a neutron star.

- The alpha decay half-life is studied in different environments and the magnitude of the screening enhancement factor is obtained. The electron clouds in the atomic environment provide screening and enhancement factor (f_α) has a value in the range of $1-2.5$. For strong magnetic field the screening enhancement factor ranges up to 10^3, 10^4 and 10^9 for $B = 10^{14,15,16}$ G respectively. Dense plasma environment having temperature of about 10^8 K with densities 10^9 g/cm^3 and 10^{10} g/cm^3 provides enhancement factor up to 10^6 and 10^9.

- The screening enhancement factor, in general, is higher for the heavy mass region as compared to the superheavy mass region, and is more prominent in certain isotopes. This may have its usefulness in studies related to chemical abundance and element formation in astrophysical sites of different conditions in the universe.

References

Abbott, B. P., Abbott, R., Abbott, T., Acernese, F., Ackley, K., Adams, C., ... others (2017). Gw170817: observation of gravitational waves from a binary neutron star inspiral. *Physical review letters*, **119**, 161101.

Arafune, J. and Fukugita, M. (1992). Electron Screening Effects on Thermonuclear Reactions in the Sun. *Progress of Theoretical Physics*, **87**, 1467-1471.

Baade, W. and Zwicky, F. (1934). Remarks on super-novae and cosmic rays. *Physical Review*, **46**, 76.

Balantekin, A., Carlson, J., Dean, D., Fuller, G., Furnstahl, R., Hjorth-Jensen, M., ... others (2014). Nuclear theory and science of the facility for rare isotope beams. *Modern Physics Letters A*, **29**, 1430010.

Banerjee, B., Constantinescu, D. H. and Rehák, P. (1974). Thomas-fermi and thomas-fermi-dirac calculations for atoms in a very strong magnetic field. *Physical Review D*, **10**, 2384–2395.

Beard, M. and Wiescher, M. (2003). The fate of matter on accreting neutron stars. *Revista Mexicana de Fisica Supplement*, **49**, 139-144.

Czerski, K., Huke, A., Biller, A., Heide, P., Hoeft, M. and Ruprecht, G. (2001). Enhancement of the electron screening effect for d+ d fusion reactions in metallic environments. *Europhysics Letters*, **54**, 449.

Frebel, A. and Beers, T. C. (2018). The formation of the heaviest elements. *Physics Today*, **71**, 1–30.

Glendenning, N. K. (1997). *Compact stars: Nuclear physics, particle physics and general relativity*. Springer New York.

Hausmann, M., Aaron, A., Amthor, A., Avilov, M., Bandura, L., Bennett, R., ... others (2013). Design of the advanced rare isotope separator aris at frib. *Nuclear Instruments and Methods in Physics Research Section B: Beam Interactions with Materials and Atoms*, **317**, 349–353.

Huang, K.-N., Aoyagi, M., Chen, M. H., Crasemann, B. and Mark, H. (1976). Neutral-atom electron binding energies from relaxed-orbital relativistic hartree-fock-slater calculations $2 \leq Z \leq 106$. *Atomic Data and Nuclear Data Tables*, **18**, 243-291.

Kadomtsev, B. B. (1970). Heavy Atom in an Ultrastrong Magnetic Field. *Soviet Journal of Experimental and Theoretical Physics*, **31**, 945.

Liolios, T. E. (2001). Multielectron sefs for nuclear reactions involved in advanced stages of stellar evolution. *Nuclear Physics A*, **693**, 847-860.

Liolios, T. E. (2003). Screened α decay in dense astrophysical plasmas and superstrong magnetic fields. *Physical Review C*, **68**, 015804.

Liu, M.-Q., Yuan, Y.-F. and Zhang, J. (2009). Effect of electron screening on the collapsing process of core-collapse supernovae. *Monthly Notices of the Royal Astronomical Society*, **400**, 815-819.

Meyer, B. S. (1989). Decompression of initially cold neutron star matter-a mechanism for the r-process. *The Astrophysical Journal*, **343**, 254–276.

Mitler, H. E. (1977). Thermonuclear ion-electron screening at all densities. I-Static solution. *The Astrophysical Journal*, **212**, 513-532.

Mussack, K. and Däppen, W. (2011). Dynamic screening correction for solar p–p reaction rates. *The Astrophysical Journal*, **729**, 96.

Potekhin, A. Y. (2010). The physics of neutron stars. *Physics-Uspekhi*, **53**, 1235–1256.

Potekhin, A. Y. and Chabrier, G. (2013). Electron screening effect on stellar thermonuclear fusion. *Contributions to Plasma Physics*, **53**, 397–405.

Rehm, K. (2016). Reactions on the surface and inside of neutron stars. In *Epj web of conferences* (Vol. 109, p. 04001).

Salpeter, E. E. (1954). Electrons Screening and Thermonuclear Reactions. *Australian Journal of Physics*, **7**, 373.

Santhosh, K., Priyanka, B. and Nithya, C. (2016). Feasibility of observing the α decay chains from isotopes of shn with Z=128, Z=126, Z=124 and Z=122. *Nuclear Physics A*, **955**, 156-180.

Schilbach, T., Trujillo, J. and Caballero, O. (2018). Nuclear physics and astronomical observations of compact objects. *Nucleus*(**63**).

Somov, B. V. (2012). *Plasma astrophysics, part 1: Fundamentals and practice* (Vol. 391). Springer Science & Business Media.

Tsuruta, S. (2016). Temperature of neutron stars. *International Journal of Modern Physics D*, **25**, 1630026.

Wan, N., Xu, C. and Ren, Z. (2015). Effects of electron screening on α-decay half-lives in different external environments. *Physical Review C*, **92**, 024301.

Wang, B., Bertulani, C. A. and Balantekin, A. B. (2011). Electron screening and its effects on big-bang nucleosynthesis. *Physical Review C*, **83**, 018801.

Zinner, N. T. (2007). Alpha decay rate enhancement in metals: An unlikely scenario. *Nuclear Physics A*, **781**, 81–87.

Printed by BoD™ in Norderstedt, Germany

9 781805 299790